MACHINERY MAINTENANCE

A practical guide using basic language to explain good machinery ma  ience

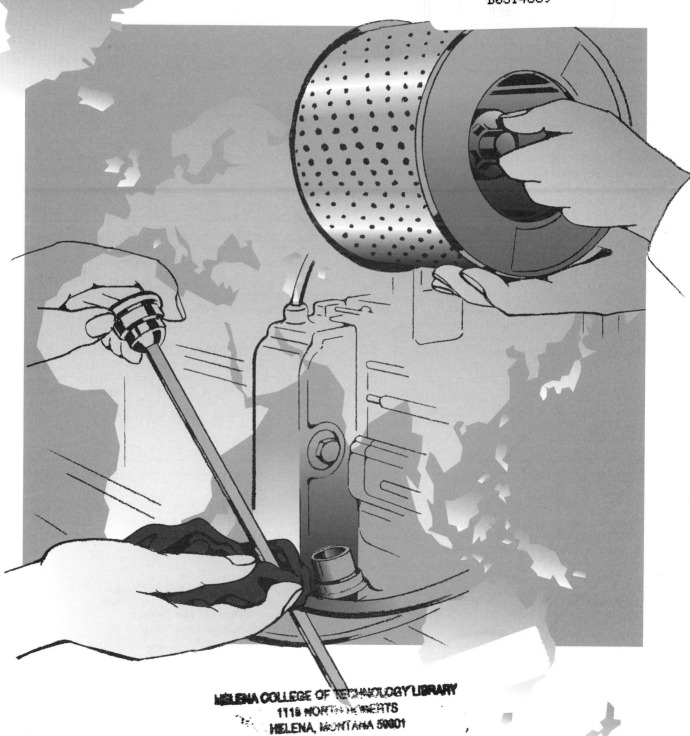

FUNDAMENTALS OF SERVICE

CONSULTING AUTHOR: Frank Buckingham, agricultural engineer and noted freelance agri-business writer who has written numerous articles and publications on agricultural machinery.

CONSULTING AUTHOR: Keith R. Carlson, experienced high school vocational agriculture instructor. Mr. Carlson is the author of many instructor's guides including the American Oil Company's "Vo-Ag Management Kits." Mr. Carlson is presently general manager of Agri-Education, Inc.

CONSULTING AUTHOR: John A. Conrads, a retired John Deere service executive, spent most of his career in overseas marketing and traveled in over 50 countries. He trained technicians and maintenance personnel in many developing nations and is the author of numerous articles on the subjects of service, training, and communication to a world-wide market place.

CONSULTING EDITOR: David M. Richardson, 10 years world travel, 54 countries, worked as a teacher in Swaziland, Malaysia, Equador, India, and the United States. Administered a bush road building camp. Organized the Amazonian Island Fishing Co-op. Presently working as a volunteer with the Peace Corps in Tahiti.

CONSULTING EDITOR: William E. Field, has traveled and worked in Hong Kong, and Guyana; done agricultural extension work dealing with pest control, grain drying and storage, and produce marketing. Published 21 articles and publications dealing with farm safety. General agricultural instructor at high school and college level. Presently extension safety specialist, Purdue University, West Lafayette, Indiana, 47907, U.S.A.

CONSULTING EDITOR: John S. Balis, has accumulated years of experience teaching farm equipment courses in the United States and India. Mr. Balis is currently working with the United States International Development Cooperation Agency.

CONSULTING EDITOR: Ron Heisner has extensive experience in agriculture in Peru and Nigeria and has traveled in Malaysia and Thailand in connection with the Future Agricultural Research Manpower project of the U.S. Peace Corps.

Mr. Heisner has written several publications including: "Bring Chunks of Reality into the Classroom," "Education for Agribusiness Occupations," and others. He has served on the Agricultural Panel at the International Education Conference, Washington, D.C. and testified before the task force for the Future of Illinois.

Mr. Heisner is now Project Supervisor for the Nigerian Manpower Project of Kishwaukee College in Illinois developing a technical agricultural program for 39 Nigerian men.

CONSULTING EDITOR: Harlow Gene Peuse gathered 18 years of practical farm experience in East Africa while working for the U.S. Peace Corps. The University of Illinois awarded him a Doctor of Philosophy in Agricultural Education. Dr. Peuse has written several pieces on his work such as "Helping Provide Food For The World's Growing Population," "The World's Chief Food Crops," "International Exchange and Work Opportunities," "Animals in World Agriculture," and others. Dr. Peuse is now working with the Agricultural Extension system in Tanzania, East Africa.

CONSULTING EDITOR: Harold Gates is an instructor at Kishwaukee College in Illinois. He spent 28 years working with engines. He is a member of the United Auto Racing Association and is an accomplished auto racer.

Mr. Gates now teaches engines and related technology to students at Kishwaukee College in Illinois. He is active in teaching in the Nigerian manpower project.

CONSULTING EDITOR: William A. Rockstroh has an accumulated 32 years of experience in technical writing, editing, and technical services management. Mr. Rockstroh is currently president of M & B Supply, a supplier of technical publications.

PUBLISHER

Fundamentals of Machine Operation (FMO) is a series of manuals created by Deere & Company. Each book in the series was conceived, researched, outlined, edited, and published by Deere & Company. Authors were selected to provide a basic technical manuscript which could be edited and written by staff editors.

PUBLISHER: DEERE & COMPANY SERVICE PUBLICATIONS, Dept. FOS/FMO, John Deere Road, Moline, Illinois 61265-8098.

SERVICE PUBLICATIONS EDITORIAL STAFF
Managing Editor: Louis R. Hathaway
Editor: John E. Kuhar
Promotions: Lori J. Dhabalt

We have a long-range interest in good machine operation

JOHN DEERE

Library of Congress Card Number: 810130

ISBN 0-86691-130-8

CONTENTS

CHAPTER 7 — ELECTRICAL SYSTEMS

CHAPTER 8 — POWER TRAINS

CHAPTER 9 — HYDRAULICS

CHAPTER 10 — INDIVIDUAL PARTS

CHAPTER 11 — SAFETY

APPENDICES

UNDERSTAND THE PURPOSE OF THIS BOOK

The goal of this text is to provide basic practical facts about machine maintenance for the world audience. It is written by experts with worldwide machinery maintenance experience.

EDITORIAL POLICY

Our basic editorial policy for this book is:

● *Stick to the facts that our advisers feel are needed.*

● *Use a basic language that can be translated into other languages with a minimum loss of meaning. Obtain the best translators available so the information can be translated with minimum distortion.*

● *Make no special efforts to inject pictures of people, machines, or landscapes from particular regions or write text in special local language idioms.*

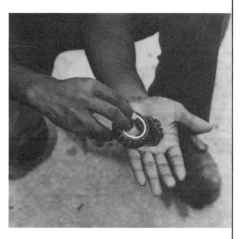

● *Stress the importance of safe maintenance practices throughout the book and provide a special section on safety as it relates to maintenance procedures.*

● *Explain maintenance practices on a tractor, since a tractor has almost all of the systems found on other engine-powered agricultural machines. It would be virtually impossible to cover each machine.*

● *NOTE: In spite of our efforts, oversights do happen. If you see any way that we can be more objective in telling about machine maintenance, please let us know.*

This book must not be used as a **replacement** for a specific machine operator's or technical manual. Always read the literature that is supplied with a particular machine for **specific** instructions.

In summary, this book was written for the international audience. So be prepared for some new and unusual words and foreign looking pictures. Do not feel the book was written for someone else when you see something foreign. It was written for everyone. It was written for you. It is natural to find unusual things in an international book. Look up strange words in the glossary. Ask your teacher about them. Learn as much as you can about them.

We, who put this book together, wish you the best of luck in your study and hope this text helps you grow and prosper.

INTRODUCTION
CHAPTER 1

INTRODUCTION

Whether you farm in Africa, Europe, America, Asia, or Australia, machines are one of your biggest concerns. It makes good sense to take good care of these expensive machines so they last long and run well. As you read this book you will learn more about this maintenance concept of caring for machines. The book is not an operator's manual for a certain machine. Rather, it is a maintenance fact book for machines in general. Also, we chose a tractor to explain good maintenance, because a tractor has almost all the systems that you will find on other engine-powered equipment.

This first chapter will help you understand:

- *What maintenance is and why it is done*
- *Some basic machine principles*
- *What needs maintenance*
- *When to do maintenance*

WHAT IS MAINTENANCE?

Maintenance is **the regular care** machines need to work well, safely and long. Keep in mind maintenance is not repairing a machine after it breaks. Maintenance is **protecting a machine so it doesn't break down or wear out too quickly.**

There are three main enemies you must protect machines from:

- **Wear** *(Grease and oil are used to protect moving parts from wear.)*

- **Dirt** *(Filters are used to catch and hold dirt before it gets inside and damages parts. Frequent, thorough washing protects the outside of the machine. Careful storage of oil, grease, and fuel and clean filters keep out dirt to protect the inside.)*

- **Heat** *(The cooling system protects the machine from heat if you make sure it has good, clean coolant, doesn't leak, and all parts are in working order.)*

WEAR

Wear is the first enemy you must protect machines from. The parts in a machine rub and turn against each other until they get hot and begin to wear. This rubbing is friction. Friction causes wear. You can feel how friction warms your hands if you twist a stick, as shown. Lubricants and bearings help protect engines from friction and thus from wear.

Lubricants

Lubricants protect machine parts from friction. You can feel how lubricants work if you put a lubricant like oil, grease, or even animal fat on your hands and twist the stick again. It will be slippery. You will feel it slip and slide, and it won't get hot because the lubricant reduces friction. When friction is reduced, wear is reduced. Friction is reduced between machine parts the same way.

There are many kinds of lubricants. Using the wrong lubricant can result in machine damage. So, be careful. Use good judgment. Only use lubricants recommended in your operator's manual.

OIL OR GREASE

Bearings

Bearings also protect machines from wear. Bearings are small metal balls, rollers, or metal cylinders that fit between moving parts; or stationary and moving parts, to support them and keep them from rubbing on each other. You can see how a roller bearing works if you get the stick again and roll it between your hands, as shown.

You can feel how the stick keeps your hands apart and turns so your hands don't rub against one another. The stick reduces friction just like a roller bearing.

But, bearings can't protect unless you protect them. Bearings must be lubricated with grease or oil or they will get hot and wear away, just as any moving part will.

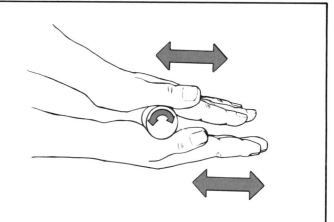

DIRT

Wear is the first enemy, and dirt is the second enemy you must protect machines from. To see the harm dirt can do, put a little dirt in the palm of your hand and twist that stick again. You will feel the dirt scratch, and your hands will heat up fast. The same thing happens if dirt gets in between machine parts. Dirt can easily scratch and wear moving parts.

Machines have filters to catch and hold dust and dirt so they can't get inside and cause this damage.

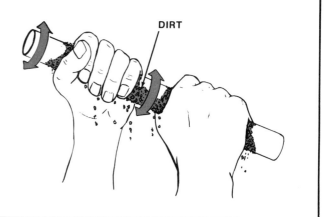

DIRT

Filters

There are filters for oil fuel, and air. Some engines also use filters for engine coolant.

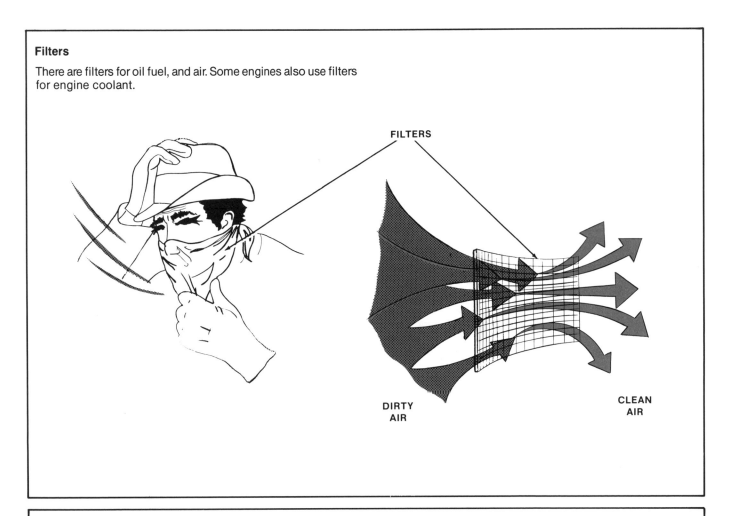

FILTERS

DIRTY AIR

CLEAN AIR

An **oil filter** catches and holds dirt that gets into the oil. Oil is pumped through the oil filter where dirt is caught in a filter screen and held so it can't go further and damage the machine.

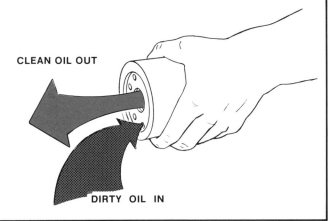

CLEAN OIL OUT

DIRTY OIL IN

Fuel filters catch dirt that gets into the fuel. They catch and hold dirt before it can get into the fuel system and damage the engine.

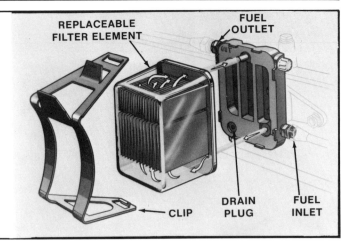

REPLACEABLE FILTER ELEMENT

FUEL OUTLET

CLIP

DRAIN PLUG

FUEL INLET

Engines pull in lots of air. The air is pulled in through an **air filter** which catches and holds dust and trash so it can't get into the engine and damage it.

HEAT

Heat is the third main enemy you must protect machines from. Heat from burning fuel and friction can heat an engine enough to make it crack or even melt. So, engines have cooling systems to control heat. You can feel how a cooling system controls heat if you twist the stick hard enough to get it warm; warm enough to be uncomfortable. Then, have a friend pour cool water over your hands while you keep turning the stick. You will feel the stick stays cool as you turn it. You have made a simple cooling system.

There are two kinds of cooling systems for engines. One cools with water, and the other cools with air.

A water-cooled engine cools itself by pumping a liquid through engine passages called the water jacket. As the coolant flows through, it absorbs heat and carries it away. Then, the coolant is pumped through a radiator that gets rid of the heat by radiating it into the air like a hot stove radiates heat.

An air-cooled engine cools itself by blowing air over its hot surfaces (cooling fins) with a fan. The blast of air carries away heat.

On some machines there are also coolers for transmission and hydraulic oil.

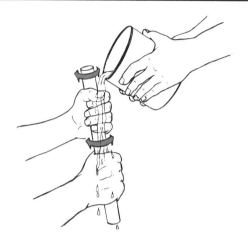

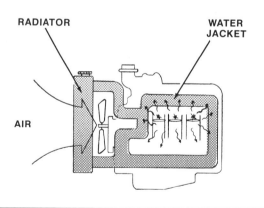

RADIATOR

WATER JACKET

AIR

WHY MAINTAIN MACHINES?

If you operate farm machines, you know time is valuable. If a job takes extra time to finish, it costs more in fuel, lubricants, wages, and machine wear. And, if a job takes extra time, other jobs are postponed.

Good maintenance keeps machines ready for work. It prevents delays and saves time, money, and crops.

Good maintenance also makes the machine more safe to operate.

WHAT NEEDS MAINTENANCE?

Each of these need regular maintenance:

- *Fuel system*
- *Air intake and exhaust system*
- *Electrical system*
- *Cooling system*
- *Engine oil system*
- *Grease points*
- *Power train (Gear boxes)*
- *Hydraulic system*
- *Tires*
- *Brakes*
- *Individual parts*

The following chapters will describe the workings and maintenance of these ten systems plus other individual parts. But first, look at these eleven short summaries. They will help you develop a clear idea of what needs maintenance, and they will prepare you for the more detailed information in other chapters.

FUEL SYSTEM

There are five main parts in an engine fuel system:

- *A fuel storage tank to store fuel*
- *Tubing to carry fuel to the engine (fuel line)*
- *Fittings to connect all the parts*
- *Filters to catch and hold dirt so it can't get into the engine*
- *A carburetor to mix fuel and air or fuel injectors to spray fuel into the engine cylinders for burning*

If dirt or water get into the fuel, they will reduce power and damage the fuel system and engine.

Regular maintenance is needed to:

- *Keep the fuel clean*
- *Keep the fuel filter clean*
- *Prevent leaks which waste fuel and may cause a fire*

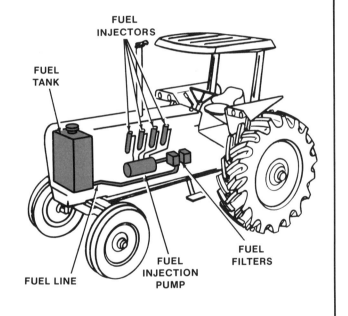

AIR INTAKE AND EXHAUST SYSTEM

An engine's intake system carries air and fuel into the engine with four main parts:

- *an **air filter and precleaner** to catch and hold dirt*
- *sometimes a **turbocharger** to increase power*
- *an **intake manifold** to carry air (diesel engine) or an air and fuel mixture (spark ignition engine) to the cylinders for burning*
- ***intake valves** to open and let air, or an air and fuel mixture, into the cylinders at the correct time*

(continued on next page)

The exhaust system carries exhaust gases away with four main parts:

- **exhaust valves,** *which open and let exhaust gas out of the cylinders at the correct time*

- *an* **exhaust manifold,** *which collects exhaust gas coming from the cylinders*

- *an* **exhaust pipe,** *which carries hot, poisonous exhaust gas and sparks safely away from the engine and the operator*

- *a* **muffler** *to quiet the loud noise of the engine exhaust*

Regular maintenance:

- *Clean or replace dry element air filters when they get dirty so only clean air goes into the engine.*

- *Empty dirty oil bath air cleaners, clean the element, and pour new oil in so it can filter dirt out of the air.*

- *Tighten fittings on the exhaust system so hot, poisonous exhaust gases don't leak out.*

- *Replace the exhaust muffler when it burns out or is damaged so that sparks are contained, and the excessive noise doesn't damage the hearing of the operator.*

ELECTRICAL SYSTEM

The electrical system has four main parts:

- *An* **alternator or generator** *to make electricity*

- **Wires and switches** *to carry electricity to lights, spark plugs, starter motor, etc.*

- *A* **battery** *to store electricity*

- **Fuses or circuit breakers** *to protect the electrical system*

Electricity can't be made if the belt from the fan pulley to the alternator or generator is loose. Electricity cannot be sent where it's needed if wires or connections are broken or loose. And, electricity cannot be stored if the battery needs water, is cracked, corroded, or worn out.

Regular maintenance:

- *Keep correct belt tension.*

- *Keep wires clean.*

- *Repair worn or broken wires.*

- *Keep battery water up to the correct level.*

- *Replace burned-out light bulbs.*

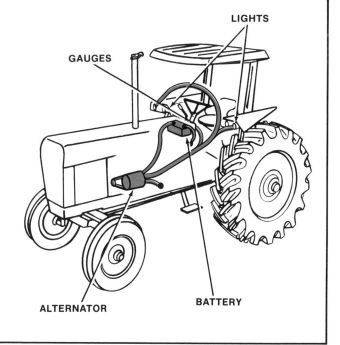

LIGHTS

GAUGES

ALTERNATOR

BATTERY

COOLING SYSTEM

Some engines are air cooled and require little maintenance, except cleaning. But, most engines are cooled by water.

A water-cooled cooling system has seven main parts:

● *A water jacket to channel water around the hot engine*

● *Hoses to connect the radiator and water jacket*

● *A radiator to get rid of the heat carried out of the engine in the water*

● *A radiator cap to hold in pressure, to allow coolant to be added, and allow overflow*

● *A water pump to move water through the system*

● *A thermostat to maintain an even coolant temperature in the cylinder block*

● *A fan to help remove heat from the radiator.*

If the cooling system gets clogged, starts leaking, or the fan slows down, or stops, the engine will overheat and damage itself.

Regular maintenance:

● *Wash insects and chaff off of the outside of the radiator.*

● *Mix the correct amount of antifreeze or rust inhibitors with the water to prevent rust, deposits, and freezing.*

● *Keep the system filled to the correct level with clean coolant.*

● *Repair leaks in the radiator, water jacket, and hoses.*

● *Make sure the fan belt has the right tension. Replace it if it is damaged.*

● *Have the thermostat and water pump checked by a mechanic to be sure they are working properly.*

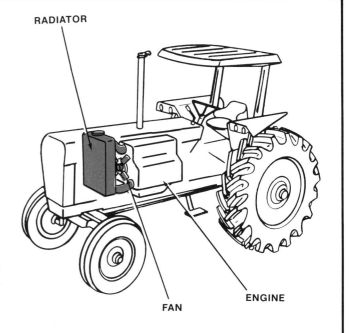

RADIATOR

FAN   ENGINE

ENGINE OIL SYSTEM

The engine oil system's job is to reduce friction, help cool the engine, create a seal between the piston rings and cylinder walls, and clean engine parts. The system has six main parts:

● *A **crankcase** to store oil*

● *An oil **pump** to provide pressure lubrication*

● ***Tubing and fittings** to circulate oil*

● ***Filters** to catch and hold dirt*

● *Oil **coolers** to cool oil*

● *The **oil** itself*

Regular maintenance:

● *Keep the right amount of oil in the system.*

● *Keep the recommended kind of oil in the system.*

● *Change the oil at recommended times.*

● *Change the filter at the recommended times.*

● *Keep fittings tight so oil doesn't leak out, and dirt doesn't get in.*

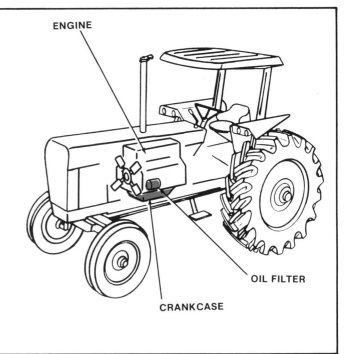

ENGINE

OIL FILTER

CRANKCASE

GREASE POINTS

Several places on a machine must be greased. For example, parts in the steering system and wheel bearings rub on one another, get hot, and wear out unless they are protected with grease. Your operator's manual will tell you what parts of your machine need greasing and when to grease them.

Front wheel bearings must usually be removed, cleaned, and packed with new grease. But, most other parts that need greasing have grease fittings. You snap the grease gun nozzle on each grease fitting and pump in a few squirts of grease.

Combines, mowers, and most all farm machines have parts that need grease. Here are some of them:

- *Front-wheel bearings*
- *Rear-wheel bearings*
- *Steering gear*
- *Clutch-throwout bearing*
- *Universal joints*
- *3-point hitches*
- *Some bearings on rotating shafts*
- *Some parts that slide against each other*

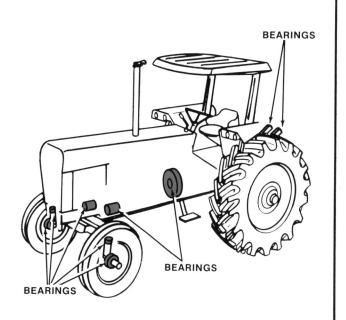

POWER TRAIN

The power train carries power from the engine to the wheels and power takeoff (PTO). It has four main parts:

- *A clutch to start and stop the power flow*
- *A transmission to change ground speeds and go forward or reverse*
- *A differential to permit drive wheels to travel at different speeds when the machine is turning.*
- *A final drive to send power to each wheel*

Some drive train parts carry very heavy loads, and some turn very fast. All must have good, clean oil.

Regular maintenance:

- *Lubricating oil for the drive train must be changed at the recommended time. Drain out the old oil and fill the drive train oil reservoir with new oil.*
- *Filters must be replaced when oil is changed so they can catch and hold rust and dirt.*

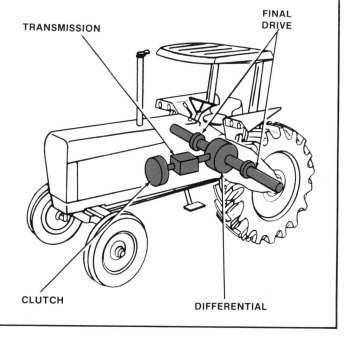

HYDRAULIC SYSTEM

Most farm machines have hydraulic systems to make steering and braking easier. shift gears, and provide the power for implements, loaders, and other equipment.

Think of hydraulics as an extension of your strength and the engine's power.

Regular maintenance:

● *Keep the hydraulic oil clean. Change it at recommended intervals.*

● *Change filters when recommended.*

● *Keep the hydraulic oil reservoir filled to the right level.*

● *Repair leaks.*

● *Keep fittings tight and clean.*

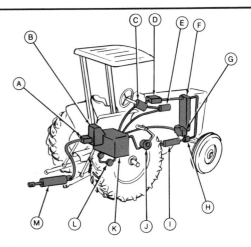

(A) REMOTE COUPLER
(B) POWER LIFT (EQUIPMENT CONTROL)
(C) POWER STEERING VALVE
(D) REMOTE CONTROL VALVE
(E) ACCUMULATOR
(F) OIL COOLER
(G) MAIN HYDRAULIC PUMP
(H) FILTER
(I) POWER STEERING CYLINDER
(J) POWER BRAKE VALVE
(K) TRANSMISSION (FLUID RESERVOIR)
(L) POWER BRAKE CYLINDERS
(M) REMOTE CONTROL CYLINDER

TIRES

Tires are designed to hold up machines and grip the ground. Too much air in them makes them ride hard, slip, and wear out fast. Too little air lets a tire slip on its rim and cut the valve stem. It can also cause the tire to heat up, buckle the sidewalls, and even make the engine work harder. The proper tire pressure is recommended in the operator's manual.

Regular maintenance:

● *Check air pressure every day.*

● *Look at tires every day to see if they have been cut or scraped.*

● *If you find damage or incorrect air pressure, repair the damage or put in the correct air pressure.*

BRAKES

Machines stop when you push the brake pedal because brake shoes are pushed against turning drums or pads squeeze turning discs. This slowly wears away the shoes or pads.

Regular maintenance:

● *Test brakes before you need them by trying to stop. Notice if the brakes pull your machine one way or the other. Have the brakes adjusted or repaired if they pull, grab, or don't function well. You and others could be injured in an accident if the brakes are faulty.*

● *When brakes wear, parts should be replaced by a mechanic.*

● *Brakes must be adjusted regularly by a mechanic. Disc brakes don't require adjustment, but worn pads must be replaced.*

 CAUTION: Brakes are very important to your personal safety and the safety of others. Maintain them well.

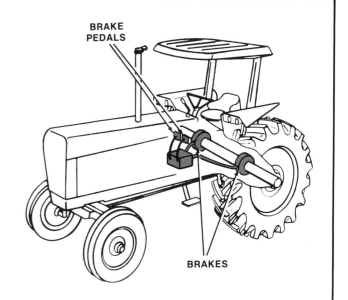

BRAKE PEDALS

BRAKES

INDIVIDUAL PARTS

● *Adjust fan belts so they have the right tension. Replace them if they are worn.*
● *Adjust chain tension. Chains need oil to keep them from wearing. They must be replaced if they are worn.*

● *Replace worn or cracked windshield wiper blades.*
● *Mirrors must be replaced when they crack or cloud.*
● *Sealed beams must be replaced when they burn out.*
● *When a light bulb burns out, a new one must be put in.*
● *Replace cut, cracked, or disconnected wires.*

WHEN SHOULD MAINTENANCE BE DONE?

Most operator's manuals recommend that you do maintenance every so many hours of engine operation. For example, every 10 hours, 100 hours, or 1,000 hours.

Most machines have an hourmeter that records the hours the engine runs. If you write down the reading when you do a maintenance job, you can always figure out when to do the job again, as follows:

Present hourmeter reading .... 40 hours

minus reading when maintenance
    was last done .... −30 hours

equals time since last maintenance .... 10 hours

(The 10-hour maintenance jobs must be done.)

Some general guidelines for maintenance times:

● *If a machine is operated 24 hours a day, do the 10-hour maintenance items twice a day.*

● *If a machine is used less than 1,000 hours each year, perform all the 1,000 hour maintenance items once a year.*

If the machine is operated in very dusty, dirty, or muddy conditions it may require more frequent maintenance.

Always refer to the specific operator's manual for complete information.

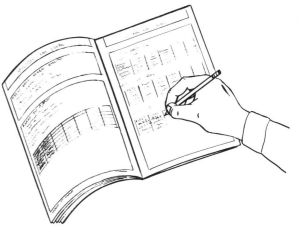

WHAT ARE SOME OF THE SIGNS THAT SAY MAINTENANCE IS NEEDED?

Besides the maintenance times recommended in your operator's manual, careful attention to the appearance, sound, feel and smell of a machine can tell you when maintenance is needed. Careful attention may avoid serious breakdowns.

LOOK

 CAUTION: Do not get close to moving parts when looking for damaged parts or listening for noises.

● **Look** *for damaged parts and watch the gauges for over-heating, reduced oil pressure, an alternator not charging, and other signs of trouble.*

● **Listen** *for unusual noises like squeaks, excessive vibration, or knocking sounds. They warn you of broken or badly worn parts or bad fuel.*

LISTEN

CAUTION: Shut off the engine and wait until all moving parts have stopped.

● **Feel** *for loose belts, chains, nuts, bolts, and unusual vibrations, but,* **never put your hands on moving belts, fans, chains and parts of the machine that can pinch or crush you.**

DO NOT FEEL FOR LEAKS IN THE COOLING, HYDRAULIC SYSTEM OR ANY COMPARABLE SYSTEM WITH LIQUID UNDER PRESSURE. Some liquids injected under your skin must be removed immediately by a surgeon who is familiar with this particular type of accident. Fluids injected under the skin can cause gangrene. (See page 102 and 103 -HOW TO FIND LEAKS).

FEEL

● **Smell** *for overheated bearings or electrical equipment, slipping belts, leaking fuel, and fires around the exhaust system.*

Remember, enclosed cabs are comfortable, but they make it more difficult to hear, smell, and feel the machine. If you have an enclosed cab, use your eyes more to make up the difference. Watch gauges carefully. Look the machine over even more frequently for signs of trouble.

SMELL

UNEXPECTED SITUATIONS

When in the field, you can't always do maintenance jobs at the exact time suggested in the operator's manual. There are unexpected situations. For example, your machine is due for an oil change, but a rain storm is rolling in and you need to finish. If your machine has been properly maintained, you can safely run an extra hour or two to get the job done before the rain comes down. You have a margin of safety with good maintenance. However, maintenance cannot be delayed very long. If maintenance is suggested at 50 hours, you may be able to go 55 hours, under good conditions. But don't delay very long. Machine wear will increase rapidly.

MACHINE MAINTENANCE AND GOOD JUDGMENT

Practice good judgment by showing it. Good judgment is shown when you use the correct fuel, oil, and grease, and do maintenance at the recommended times. Correct decisions prevent machine breakdowns and also help avoid accidents. For example, if you test your brakes before you need them, you have used good judgment and reduced the chance of an accident. It may even save your life or the life of others.

 CAUTION: Machines are very powerful. Respect them! Always follow the safety instructions in the machine operator's manual and on safety signs that are on the machine. This can keep you from getting injured or killed.

GET TO KNOW YOUR MACHINE LIKE A FRIEND

Get to know your machine. Grease points, oil dipsticks, batteries, oil filters, and other items may be in different places on different machines. Your operator's manual will show you where to find them. So, take time to read it and learn as much as you can about your machine. The more you know about your machine the easier you will be able to perform maintenance work. You will also be able to operate the machine better and more safely.

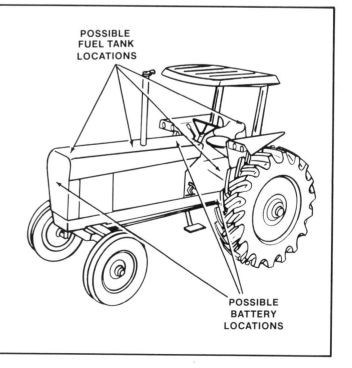

POSSIBLE FUEL TANK LOCATIONS

POSSIBLE BATTERY LOCATIONS

USE THE RIGHT NAME

Most machines have an alternator to make electricity. But, some have a generator. Some engines have dry element air filters, others have oil bath filters. There are many parts. In different parts of the world the same part may even have different names. The point is, things are confusing enough. So, please call parts by the name used in your operator's manual, so you don't add to the confusion.

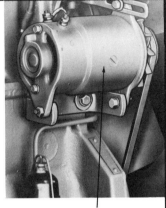

ALTERNATOR   **GENERATOR**

BE PROUD OF GOOD MAINTENANCE AND SAFETY PRACTICES

You can be proud of a well maintained machine. Good maintenance and careful operation make it look better and perform better. It displays your good judgment and saves money by being more efficient. If you drive a smooth running machine you won't get tired as quickly and you will get the job done faster. These facts should be goals for every machine operator. Keep them in mind and you will make them happen.

Above all, operate the machine safely. It prevents damage to the machine and reduces the possibility of injury or death.

SUMMARY

Maintenance is the regular care that machines need to work well, safely, and long. To give machines proper care you must protect them from their three main enemies: wear, dirt, and heat.

Oil, grease, and bearings will protect machines from wear if you make sure the machine has the right kind and amount of oil and grease, and that bearings work properly.

Filters will protect machines from dirt in the fuel, oil, and air if you make sure the right kind of clean filters are on the machine.

The cooling system will protect machines from heat if you make sure the system is clean and working, has the right water and chemical mixture, and doesn't leak.

You must learn how to care for ten main systems and some individual parts in machines.

- *Fuel system*
- *Air intake and exhaust system*
- *Electrical system*
- *Cooling system*
- *Engine oil system*
- *Grease points*
- *Power train*
- *Hydraulic system*
- *Tires*
- *Brakes*
- *Individual parts*

However, it will do you little good to learn about systems and parts unless you also develop good maintenance judgment. Your good judgment is the most important part of maintenance. You can avoid breakdowns, delays, and frustration if you practice good judgment in your daily maintenance.

DO YOU REMEMBER?

1. What are the three main problems which make a machine fail?

a.

b.

c.

2. Metal balls, rollers, or inserts between moving parts are called _____.

3. What are the two kinds of cooling systems?

a.

b.

4. A storage tank, tubing, filters, and a carburetor are all part of a _____ system.

5. What does a radiator do?

6. A clutch, transmission, and final drive are part of a ____

_____

7. What do you do when it is time for a maintenance job and you are out in the field?

8. Why does good maintenance make you feel good?

ENGINES
CHAPTER 2

INTRODUCTION

Machines have several components. One is the engine.

This chapter will tell you:

- *How an engine makes power by burning fuel and air*
- *The basic parts of an engine and how they work together*
- *The four steps in the operation of a 4-stroke-cycle engine*
- *The difference between diesel and spark ignition engines*
- *How power from an engine is measured*

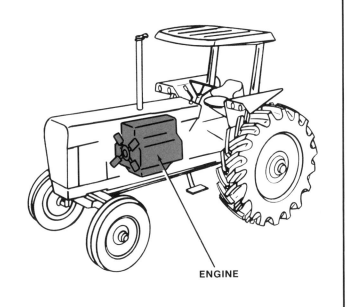

ENGINE

HOW AN INTERNAL COMBUSTION ENGINE MAKES POWER

Internal combustion engines compress air and fuel in their cylinders and burn the mixture. The hot, burning gases expand and push the piston down.

Each piston is connected to the engine's crankshaft so when it is forced down, it turns the crankshaft. Power to drive machines is taken from the turning crankshaft.

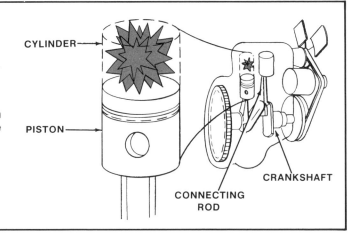

CYLINDER

PISTON

CRANKSHAFT

CONNECTING ROD

THE FIVE BASIC PARTS OF AN ENGINE

- **Cylinders** *are containers that hold the burning fuel and pistons.*
- **Pistons** *move up and down in the cylinders just as your legs pump up and down on bicycle pedals.*
- **Connecting rods** *connect the moving pistons to the crankshaft and turn the crankshaft, just as the crank on a bicycle connects the force of your feet to the crankshaft.*
- *The* **crankshaft** *turns each time a piston pushes a connecting rod down, just as the crankshaft on a bicycle turns when you pedal.*
- **Flywheel** *is a heavy wheel on the end of the crankshaft that spins with the crankshaft and helps keep the engine running with its momentum.*

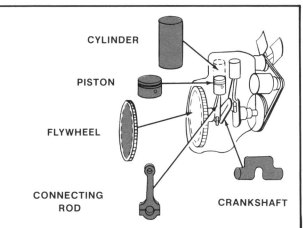

CYLINDER

PISTON

FLYWHEEL

CONNECTING ROD

CRANKSHAFT

(continued on next page)

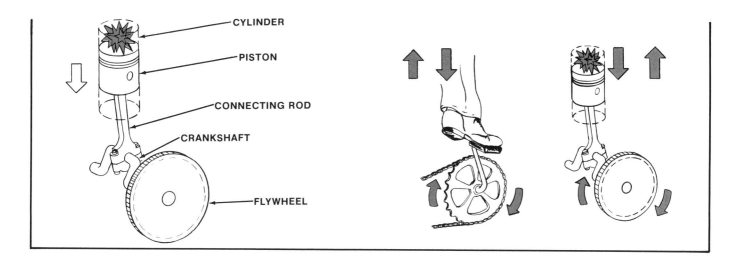

Each time a piston moves down or up, it makes one stroke. Four-stroke engines must make four strokes to complete one cycle. It takes 2 revolutions of the crankshaft to complete one cycle.

The four strokes are:

DIESEL ENGINES

1. Intake Stroke:
   The piston is pulled down by the crankshaft and draws in **air** through a valve which opens In the top of the cylinder.

2. Compression Stroke:
   The piston is pushed back up by the crankshaft to compress and heat the **air** at the top of the cylinder.

3. Power Stroke:
   Diesel fuel is sprayed into the cylinder where it's **ignited by the hot air.** The expansion force pushes the piston down the cylinder and turns the crankshaft.

4. Exhaust Stroke:
   The piston is pushed back up by the crankshaft and pushes the exhaust gases out a valve which opens in the cylinder.

Two-stroke engines are used in small power tools, such as chain saws, mowers, and sprayers. All four steps: intake, compression, power, and exhaust happen in two strokes of the piston instead of four. It takes one revolution of the crankshaft to complete one cycle. Most two-stroke gasoline engines don't have an oil reservoir in the crankcase, so oil must be mixed with the fuel. However, diesel two-stroke-cycle engines do have an oil reservoir in the crankcase.

SPARK-IGNITION ENGINES

1. Intake Stroke:
   The piston is pulled down by the crankshaft and draws in **both air and fuel** through a valve which opens in the top of the cylinder.

2. Compression Stroke:
   The piston is pushed back up by the crankshaft to compress **both air and fuel** in the top of the cylinder.

3. Power Stroke:
   A **spark plug sparks and ignites the fuel and air** mixture in the cylinder. The expansion force pushes the piston down the cylinder and turns the crankshaft.

4. Exhaust Stroke:
   The piston is pushed back up by the crankshaft and pushes the exhaust gases out a valve which opens in the top of the cylinder.

(continued on next page)

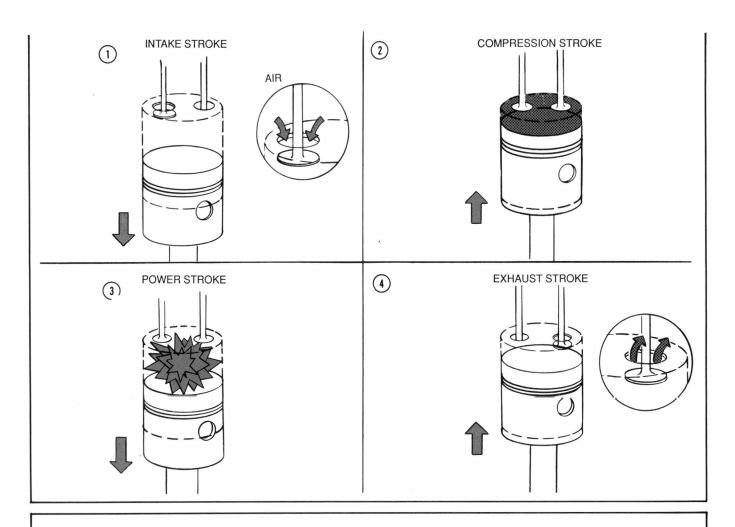

1. INTAKE STROKE — AIR

2. COMPRESSION STROKE

3. POWER STROKE

4. EXHAUST STROKE

THE DIFFERENCE BETWEEN DIESEL ENGINES AND SPARK-IGNITION ENGINES

Ignition in Spark-Ignition Engines

Fuel is mixed with air **before** it enters the cylinders. Then, the mixture is ignited by an electric spark from a spark plug in the top of each cylinder.

Ignition in Diesels

Diesel engines do not have spark plugs to ignite fuel. Instead, they compress air in their cylinders until it is hot enough to ignite fuel. When diesel fuel is sprayed in, it ignites on contact with the hot air.

Fuel

Gasoline (Petrol) is the most common fuel for spark-ignition engines. But, some burn liquified petroleum gas (LPG), natural gas, alcohol, or fuel mixtures such as gasohol (gasoline and alcohol). There have been some experiments that show **diesel** engines can burn special fuels like mixtures of diesel and sunflower or soybean oil. But don't try this kind of fuel unless it is approved by the engine manufacturer.

Engines are often identified by the type of fuel they burn.

- *Diesel engines*

- *Gasoline (Petrol) engines*

- *LPG engines.*

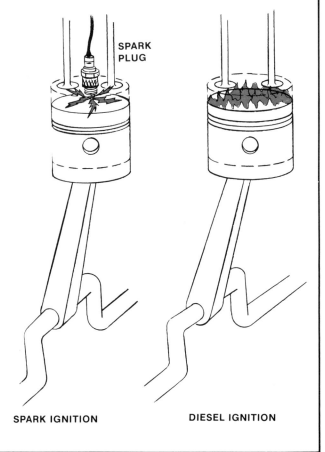

SPARK PLUG

SPARK IGNITION   DIESEL IGNITION

WHAT IS POWER?

The purpose of an engine is to make power. Power is defined as the rate of doing work. A tractor pulling a wagon up a hill at 5 miles per hour (8 kilometers per hour) must produce twice as much power as a tractor pulling the same load up the hill at 2.5 miles per hour (4 kilometers per hour).

In the past, power was expressed in horsepower (hp). But you will see power ratings in kilowatts (kW) which is the S.I. metric expression of power.

HOW IS POWER MEASURED?

Engine Power

Basically, engine power is the power an engine can develop by itself. It is measured either in horsepower or kilowatts.

Drawbar Power

Drawbar power is how much weight a tractor can pull at a certain speed.

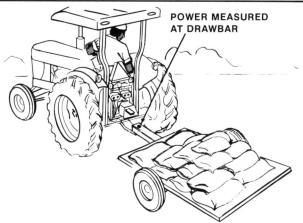

POWER MEASURED AT DRAWBAR

PTO Power

Power take-off (PTO) power is the power in horsepower (kilowatts) a tractor delivers at the PTO shaft.

Hydraulic Power

Hydraulic power is the power measured in horsepower or kilowatts that is available to drive hydraulic motors or other equipment.

Hydraulic "power" is also measured as force such as on a 3-point hitch or a hydraulic cylinder. The force is measured in newtons (N) or pounds force (lbs/f).

PTO POWER IS MEASURED HERE

COMPARING POWER

When you compare the power ratings of machines, be sure the figures are for the same kind of power. If you try to compare engine power with drawbar power, you make a wrong comparison and end up with wrong results.

Rated power may differ in different countries, because of different test conditions and procedures. Sometimes different correction factors are used. When comparing any power ratings you must carefully study how the ratings were achieved.

SUMMARY

You should understand that an engine compresses and burns air and fuel in its cylinders to make power. The expansion force of the burning fuel is turned into useful power by a set of basic engine parts working together. These basic parts are the:

- *Cylinders*
- *Pistons*
- *Connecting rods*
- *Crankshaft*
- *Flywheel*

There are four steps in an engine's operation called **strokes:**

- *Intake*
- *Compression*
- *Power (Ignition)*
- *Exhaust*

Understand that a **diesel engine** has no spark plugs. Diesel engines compress air in their cylinders until it is hot, then inject diesel fuel which ignites on contact with the hot air.

Spark-ignition engines mix fuel with air in a carburetor, **before** it enters the cylinders, and spark plugs ignite the mixture with an **electric spark.**

Some spark-ignition engines have fuel injection instead of a carburetor. The gasoline (petrol) fuel is injected into the intake port of the cylinder where it is mixed with the intake air.

Power is the rate of doing work. It is how fast a machine can pull a certain load over a certain distance. Tractor power can be measured in four different ways:

- *Engine*
- *Drawbar*
- *PTO*
- *Hydraulic*

Power figures for the same machine will be different at each of these places.

DO YOU REMEMBER?

1. How does an engine produce power?

2. Name the five basic parts that make an engine work.

3. What are the four steps in a four-cycle engine's cycle?

4. Explain the difference between ignition of gasoline (petrol) and the ignition of diesel fuel in engines.

5. What term is used to describe engine power?

ENGINE INTAKE AND
EXHAUST SYSTEMS
CHAPTER 3

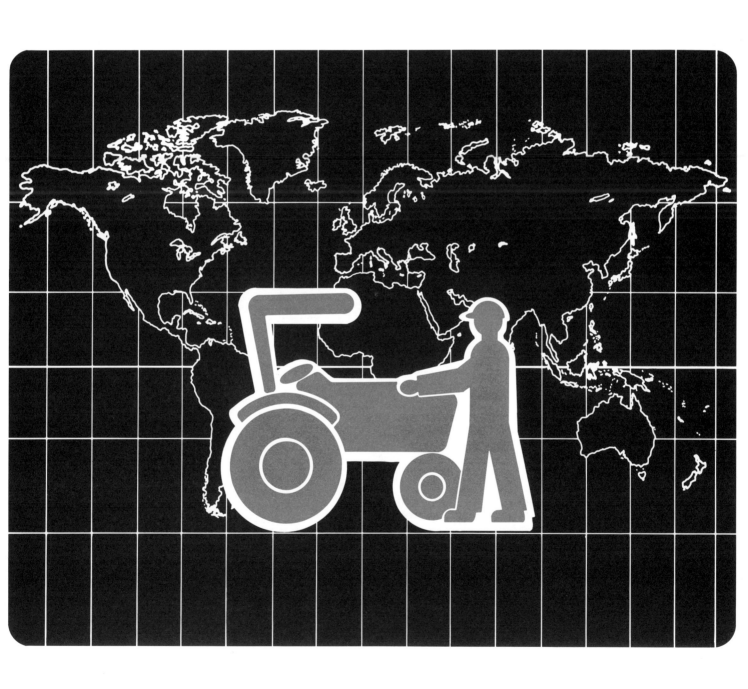

INTRODUCTION

This chapter will:

- *Describe engine intake and exhaust systems*
- *Tell what they do*
- *Describe the parts and how they work together*
- *Explain general maintenance*

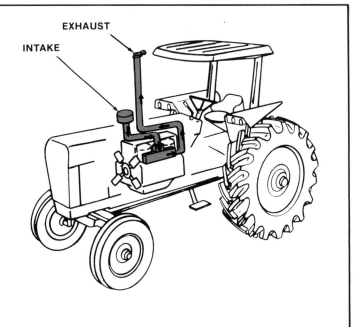

THE AIR INTAKE SYSTEM

An engine air intake system is an arrangement of tubes, filters, and fittings that carries air (oxygen) into the engine so that the fuel can be burned in the cylinders.

ENGINE INTAKE SYSTEM PARTS

An engine intake system has the following parts:

- **Pre-cleaner** *(on some engines) to catch and hold large pieces of dirt, chaff, and lint in incoming air, so they won't get into the engine and cause damage.*

- **Air cleaner** *to catch small pieces of dirt, chaff, and lint in incoming air, so they won't get into the engine and cause damage.*

- **Turbocharger** *(on some engines) to push more air into cylinders for more efficient fuel burning.*

- **Carburetor** *(on spark-ignition engines only) to mix fuel with air before it enters the cylinder.*

- **Intake manifold** *to carry air, or a fuel and air mixture, to the engine cylinders.*

- **Intake valves** *to open, at just the right time, to let air or an air and fuel mixture enter each cylinder, and to close again, at the right time; and to hold the gases in during the compression, power, and exhaust strokes.*

Pre-cleaners, pre-screeners, and filters are especially **important to machine operators** because they require almost daily maintenance.

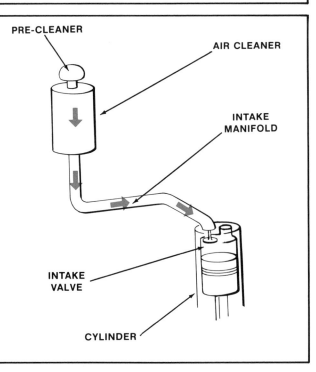

PRE-CLEANERS AND PRE-SCREENERS CATCH DIRT AND TRASH

IMPORTANT: Clean air is extremely important for long engine life.

Pre-cleaners catch some of the dirt and trash in incoming air before it can be drawn into the main air cleaner.

Some intake systems also have a pre-screener for extra protection. Pre-screeners catch lint and heavy trash before it gets to the pre-cleaner.

There are two kinds of air cleaners:

- *dry element*
- *oil bath*

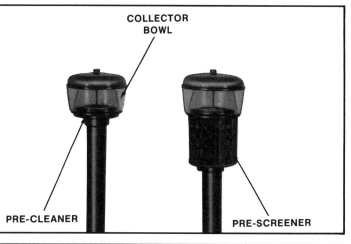

COLLECTOR BOWL

PRE-CLEANER

PRE-SCREENER

DRY ELEMENT AIR CLEANERS

Dry element air cleaners clean air two ways.

- *First, they swirl air so dust drops into a dust cup or rubber dust unloading valve where it can be emptied out.*
- *Second, they filter the dust and grit out of the air by passing the air through a paper filter that catches and holds dust.*

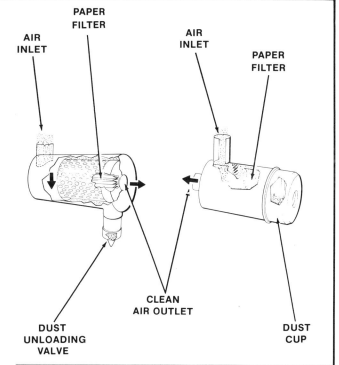

AIR INLET

PAPER FILTER

AIR INLET

PAPER FILTER

DUST UNLOADING VALVE

CLEAN AIR OUTLET

DUST CUP

Some dry element air filters have a second filter, called a secondary filter, that catches tiny dust particles that squeeze through the main filter.

DUST UNLOADING VALVE

23

OIL BATH AIR CLEANERS

Oil bath air cleaners give air a bath in oil and filter it. Incoming air is drawn down through the filter housing. The air then bubbles through oil and is forced through a wire mesh filter. Dirt and trash are caught and held in the filter where they can be cleaned out later.

 CAUTION: Oil bath cleaners must not be used on diesel engines. In certain unusual conditions, oil may be sucked from the filter and burned in the engine. This can cause uncontrolled overspeeding of the engine, possible engine damage, and personal injury.

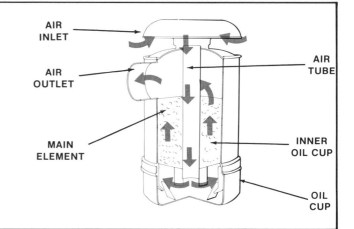

AIR FILTER RESTRICTION INDICATORS WARN YOU

Some engines with dry element air cleaners have a **filter restriction indicator** which tells you when to clean or replace the air filter.

Some restriction indicators turn on an instrument panel light when the filter is full. Others have dials or gauges beside the filter to tell you that the filter must be cleaned.

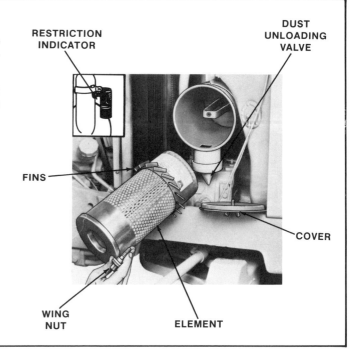

TURBOCHARGERS INCREASE POWER

Turbochargers do not clean or filter air; they increase power.

Turbochargers make more power because they push extra air into the cylinders to burn the fuel better; like blowing air on a fire.

IMPORTANT: Always let a turbocharged engine idle down to normal idle speed before you shut it off. You will wear out the turbocharger quickly if you shut off the engine at high speed because the turbocharger bearings will run out of oil before they stop spinning.

Some engines also have an intercooler between the turbocharger and the engine to cool the intake air.

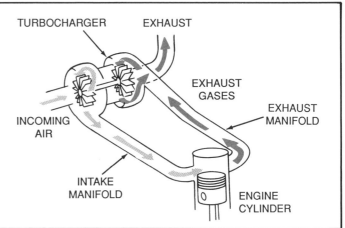

THE ENGINE EXHAUST SYSTEM

An exhaust system is an arrangement of pipes, fittings, and a muffler. It must safely carry noise and hot, poisonous gases away from the engine and operator and catch sparks that could start fires.

Engine **exhaust systems** have the following parts:

- **Exhaust valves** *to open and release exhaust gases from each cylinder, then close tightly.*

- **Engine manifold** *to collect exhaust gases.*

- **Exhaust pipe** *to pipe the hot, poisonous exhaust gases away from the engine and operator.*

- **Muffler** *to reduce hearing-damaging noise and sparks.*

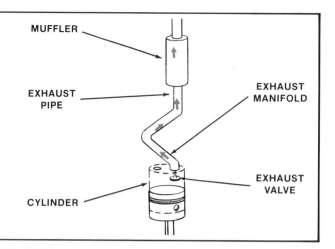

MAINTENANCE OF INTAKE AND EXHAUST SYSTEMS

The intake and exhaust systems need the following maintenance:

EMPTY THE PRECLEANER

Empty the precleaner every day.

Any time chaff covers the precleaner, stop the engine and wipe the precleaner clean. Chaff on the outside can restrict air flow and reduce power.

CLEAN OUT THE DUST UNLOADING VALVE OR DUST CUP

IMPORTANT: Always shut off the engine before you clean the dust unloading valve so unfiltered air isn't drawn into the engine.

Squeeze open the rubber dust unloading valve every day and wipe it clean. Be sure it opens freely. Replace it if it is torn or broken.

Do not operate the engine if the valve stays open. An open valve lets in unfiltered air.

Remove and wipe out the dust cup if your machine has one on the filter. Replace the dust cup if it is cracked.

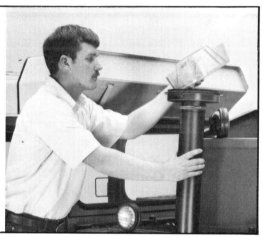

WHEN TO CLEAN OR REPLACE DRY ELEMENT AIR FILTERS

Clean or replace the filters, at intervals recommended in the operator's manual or when the filter restriction indicator shows the filter is dirty.

Clean or replace the filter if the engine starts smoking more than usual or loses power. A dirty filter reduces power.

Don't service a dry element air cleaner more often than is actually required.

HOW TO CLEAN DRY ELEMENT AIR FILTERS

Remove the filter element. Be careful not to dent the filter's screen or gasket.

Hold the filter in one hand and tap it gently against your other hand. Turn the filter slowly and tap out the dirt. Do not pound or drop the filter.

If you have compressed air, **blow** the dirt **out** of the filter from the inside.

 CAUTION: Never blow compressed air against the outside of the filter. And, do not use air pressure above 200 kPa (30 psi). High pressure could cause injury and it will blow a hole in the filter paper.

If compressed air isn't available, flush it, from the inside, with a gentle stream of water. Be careful not to tear the paper element.

When you are through flushing, set the element aside to dry. It may take a day or more to completely dry the element. So, keep a spare filter on hand to use while the other dries.

TAPPING THE ELEMENT	WASHING THE ELEMENT
BLOWING OUT THE DIRT	RINSING THE ELEMENT

WASHING CAN DAMAGE FILTERS

• *Never wash a dry element air filter in gasoline (petrol), diesel fuel, or solvent. They will destroy the filter element.*

• *Never put oil in a dry element air filter. The oil will block the air flow.*

• *After washing, never use compressed air to **dry** a wet filter element. You may blow a hole in the moist paper.*

• *Do not let a wet filter element freeze. The paper will crack.*

• *Do not dry a wet element in an oven or near a stove. The heat will weaken the filter paper.*

• *Always replace the filter element if it gets oily or smoked from the engine backfiring.*

• *Do not clean filters unless they need it. There is always a danger of damaging a filter when it is cleaned. Even a pinhole can let in enough dirt to ruin an engine.*

IMPORTANT: Even the smallest hole in a filter can cause major damage to the engine.

INSPECT DRY ELEMENT FILTER AFTER CLEANING

Stick a flashlight inside so the light shines out through the element. Pinholes, cracks, and worn places show up clearly when the light shines through them. Even a very small hole can let enough dirt in to ruin an engine. If you spot a hole, replace the filter.

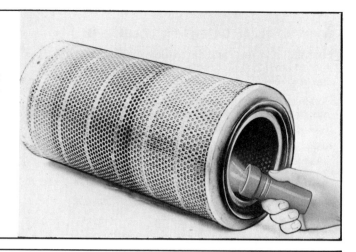

CHECK THE SECONDARY FILTER ELEMENT TOO

While the main filter is removed, look for dust on the secondary filter element. A dirty secondary filter usually means the main filter is leaking. Replace both the main filter and the secondary filter if you see dirt on the secondary filter.

Replace the secondary filter at least once a year even if it is clean, and **never try to clean the secondary filter.** It is made of different material than the main filter and cannot be cleaned or washed.

SECONDARY
FILTER
ELEMENT

HOW TO REPLACE A DRY ELEMENT AIR FILTER

If your machine has an air filter element that cannot be washed or if the element is damaged, replace it with a new one.

To replace a dry element:

1. **Remove the filter cap.**

2. **Slide out the old filter element.**

3. **Wipe out the inside of the air cleaner container with a clean, damp cloth.**

4. **Look for holes, cracks, and leaks in the air cleaner container, air tubes, and connections. Repair or replace if necessary.**

5. **Slide in a new filter element.**

6. **Be sure the filter gasket fits tight when the element is installed.**

7. **Tighten the filter cover securely.**

WHEN TO CLEAN OIL BATH AIR CLEANERS

Look at the oil level and dirt in the oil cup **every day.** If there is about 1 cm (1/2 inch) of dirt in the cup, dump it out, wipe out the cup, and refill it.

Sometimes dirt is held up in the filter and will not settle down into the oil cup. If this happens, you can still tell when to clean the cup and change oil because the oil will **feel** thick and gritty.

Once a year, about every 1,000 hours, take the filter apart and clean it thoroughly.

- *Wipe out the intake tube with a cloth soaked in diesel fuel.*
- *Wash the filter element in diesel fuel or solvent.*

 CAUTION: If the oil gets too dirty, and stays up in the filter, it may be drawn into the engine and burned. If a diesel engine starts burning oil from its air filter, there is no way to stop the engine. It could run out of control until all the oil in the air filter is burned. This is dangerous and could damage the engine.

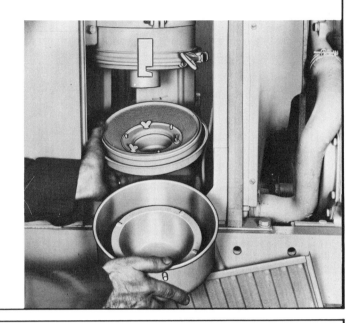

HOW TO CLEAN AN OIL BATH AIR CLEANER

 CAUTION: Do not use gasoline (petrol) for cleaning filters. It can ignite and cause serious burn injuries.

1. Remove the oil cup and wash it with diesel fuel, kerosene, or an approved solvent.

2. If there is any water in the cup, check for a leak in the air tube. Replace the tube if it leaks. If the air intake cap has come off, replace it.

3. Wipe the cup clean with a dry cloth.

4. If there is a tray or screen above the cup, clean it, too.

1. CLEAN OIL CUP IN DIESEL FUEL   2. REMOVE DIRT FROM TRAY (IF USED)

REFILLING THE OIL BATH AIR CLEANER OIL CUP

Fill the cup up to the "oil level" mark. Most manufacturers recommend using the same oil used in the engine crankcase.

Do not overfill the cup. Overfilling reduces air flow and engine power. And, excess oil may be drawn into the engine and burned. See **CAUTION** above.

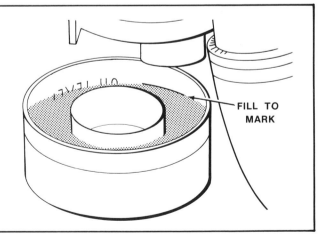

FILL TO MARK

CHECK ALL THE AIR INTAKE AND EXHAUST CONNECTIONS FOR LEAKS

Look for leaks every 500 hours. Do it more often if suggested in the operator's manual. Look for leaks in tubes and hoses, and replace the parts that leak. Make sure all the clamps and fittings are tight.

TURBOCHARGER   AIR CLEANER

CHECK INTAKE AND EXHAUST MANIFOLDS FOR LEAKS

Check the intake and exhaust manifolds for leaks every 50 hours. With the engine running, spray a small amount of oil on places you suspect are leaking. The oil will be drawn into leaks in the intake manifold or will bubble at leaks in the exhaust manifold.

Tighten loose bolts and have cracks repaired immediately. Leaks in the intake manifold let dirty, unfiltered air into the engine and cause rapid engine wear. Leaks in the exhaust manifold or muffler let poisonous exhaust gases get to the operator.

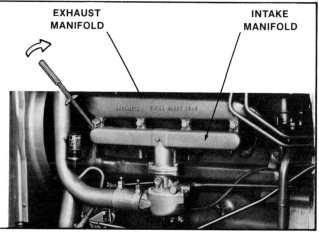

EXHAUST MANIFOLD   INTAKE MANIFOLD

ENGINE EXHAUST GAS IS DANGEROUS

 CAUTION: Never operate an engine inside a closed building unless the exhaust gas is piped outside. Open the doors and windows. The exhaust gases of all engines contain poisonous carbon monoxide. This deadly gas has no smell and no color. It can kill you.

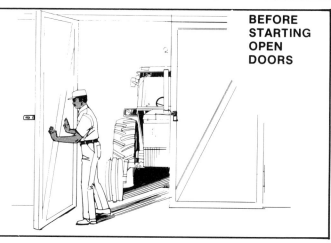

BEFORE STARTING OPEN DOORS

SUMMARY

Engines have an intake system to pull in fresh air. It is an arrangement of tubes, filters, and fittings that draws in and filters air for the engine.

Engines also have exhaust systems. They safely carry poisonous exhaust gases and noise away from the engine and operator, and catch dangerous sparks.

Filters in the intake system must be replaced or cleaned frequently. Filters are very important. Even a few specks of dust getting by these filters can scratch and wear out the parts inside the engine.

(continued on next page)

DO YOU REMEMBER?

1. To trap and hold small pieces of dirt, chaff, and lint an _____is made part of the air intake system.

2. Air can pass through paper.   True   False

3. What is used on some engines to push extra air into the cylinder?

4. Name two parts of the exhaust system.

a.

b.

5. How do you perform daily service to a dust unloading valve?

6. Why will holes or cracks in the air cleaner cause problems?

7. What problems are caused by leaks in the exhaust system?

8. What should you do if you run an engine inside?

ENGINE FUEL
CHAPTER 4

INTRODUCTION

This chapter will describe:

- *Different kinds of fuel*
- *Engine "knock"*
- *Fuel storage*
- *How to refuel engines safely*
- *Fuel systems for spark ignition, Diesel, and LPG engines*
- *Fuel system maintenance*

DIFFERENT KINDS OF FUEL

Fuel is the burnable liquid or gas that an engine burns in the cylinders to make power. Fuel is "food" to an engine. Without it, an engine cannot produce power.

There are different kinds of fuel:

- *Diesel*
- *Gasoline (Petrol)*
- *Liquified Petroleum Gas*
- *Gasohol and Diesohol*

Burning the wrong fuel in an engine will ruin it. For example, if you try to burn gasoline (petrol) in a Diesel engine, it will burn too fast, ruin the injection pump, and damage the fuel injectors.

If you do not know which kind of fuel to use, read your operator's manual, look for a sign by the fuel filler cap, or ask a dealer or mechanic.

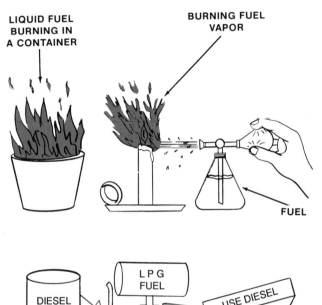

LIQUID FUEL BURNING IN A CONTAINER

BURNING FUEL VAPOR

FUEL

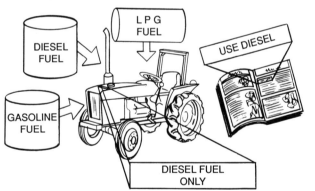

DIESEL FUEL

LPG FUEL

USE DIESEL

GASOLINE FUEL

DIESEL FUEL ONLY

DIESEL FUEL

In the United States there are two grades of diesel fuel: D-1 and D-2. D-1 fuel is for cold weather use. D-2 fuel is for warm weather use. In many other parts of the world diesel fuel is just called winter or summer fuel. If you have some winter fuel left over in the spring you can burn it. But, don't try to burn summer fuel in the winter, because it will plug your fuel filters. Plugging is caused by parafin in the fuel that crystallizes at cold temperatures. Also, look for the Cetane rating of the fuel. It should have a Cetane rating of at least 40. This assures good starting.

DIESEL FUEL GRADES

① IF YOUR ENGINE DOES THIS KIND OF WORK	② AT THIS TEMPERATURE	③ USE THIS GRADE OF DIESEL FUEL
Light load, low speed, considerable idling.	Above 27°C (80° F)	2-D
	Below 27°C (80° F)	1-D
Intermediate and heavy load, high speed, minimum of idling.	Above 4° C (40° F)	2-D
	Below 4° C (40° F)	1-D
At altitudes above 5,000 feet (1525 m)	All	1-D

GASOLINE (PETROL)

Gasoline is graded by octane numbers. Most agricultural gasoline engines burn gasoline with octane ratings of 84 to 94; often called "regular" gasoline.

Some gasoline has lead added so it will burn evenly. Unleaded gasoline has different chemicals added so it will burn evenly.

IMPORTANT: Never use unleaded gasoline in an engine designed to burn leaded gasoline. The valves may stick and burn, and the valves seats will wear out more quickly.

The engine operator's manual and labels will tell you whether to use leaded or unleaded gasoline.

GASOLINE GRADE	APPROXIMATE OCTANE NUMBER
Premium Grade	100
Regular Grade	88-94
Low Grade	70 - 75

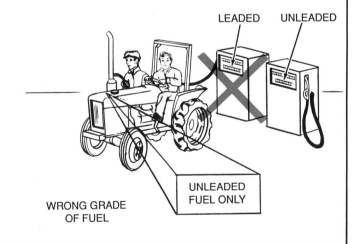

LEADED   UNLEADED

UNLEADED FUEL ONLY

WRONG GRADE OF FUEL

LIQUID PETROLEUM GAS (LPG)

LPG, often called LP-gas, is propane and butane gas mixed together. It is mostly propane. There is no standard international rating system for LPG. Some LPG has a little extra propane in it so it will burn better in cold weather. But, that is about the only difference.

GASOHOL AND DIESOHOL

Some people burn a mixture of gasoline and alcohol called gasohol or a diesel and alcohol mixture called diesohol in engines. These fuels burn well in some engines. However, there are some facts you should consider before burning it:

● *Gasohol and diesohol evaporate quickly, increasing the amount of fuel wasted in storage.*

● *Gasohol and diesohol burn easier, so there is a slightly greater danger of fire in the fueling areas.*

● *Gasohol and diesohol may corrode certain metal fuel system parts.*

● *The alcohol is a solvent. It may damage some rubber and plastic parts in the engine.*

● *These fuels often burn at a higher temperature than the engine is designed for.*

If you plan to burn gasohol or diesohol, you would be wise to discuss your plans with a mechanic or dealer.

WHAT IS ENGINE "KNOCK?"

Burning the wrong grade of diesel fuel or gasoline can make an engine "knock." You can easily hear the unnaturally loud knocking sound. It comes from fuel burning too fast, too early, or unevenly in the cylinders.

Knocking is annoying. But worse, knocking reduces power and damages valves, pistons, and bearings. It will wear out an engine fast.

If an engine starts knocking, slow down and make sure the cooling system is working properly. If the knocking continues, use a higher grade fuel. If the knocking continues, stop the engine and have a mechanic look at it. There may be a fuel injection problem, or the timing of the engine may be incorrect.

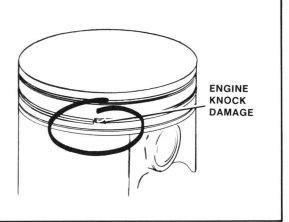

ENGINE
KNOCK
DAMAGE

STORE FUEL PROPERLY

There are three reasons for storing fuel properly:

● *Safety*

● *Protection from contamination*

● *Protection from evaporation*

FRESH
AIR

10 - 15 m   (40 ft)

FIRE
EXTINGUISHER

NO WEEDS

SAFETY

 CAUTION: Handle fuel carefully. Do not smoke while you are near fuel tanks. Also keep sparks and flames away from fuel tanks.

Fuel burns easily, especially gasoline (petrol) and LPG. Spilled fuel and fumes can be ignited by a single spark or flame. Diesel fuel will burn too if a flame touches it or if it is spilled on a hot engine. For safe storage:

- *Keep fuel storage tanks away from buildings.*
- *Don't spill fuel.*

- *Don't let explosive fumes collect in closed buildings or low places.*
- *Mark LPG, gasoline, and diesel fuel tanks clearly with signs so people know what they are.*
- *Clean weeds and trash away from fuel storage areas so accidental fires cannot spread.*
- *Keep a fire extinguisher handy to put out fires.*

CONTAMINATION

Store fuel so dirt and water don't collect. Clean fuel makes engines easier to start and reduces fuel system maintenance. Dirt and water in fuel can clog small passages, like diesel injectors, and stop an engine.

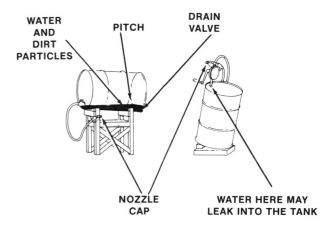

WATER AND DIRT PARTICLES — PITCH — DRAIN VALVE — NOZZLE CAP — WATER HERE MAY LEAK INTO THE TANK

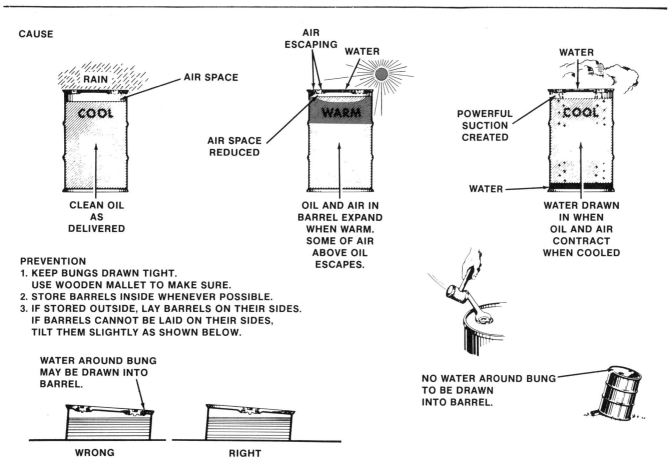

CAUSE

CLEAN OIL AS DELIVERED

OIL AND AIR IN BARREL EXPAND WHEN WARM. SOME OF AIR ABOVE OIL ESCAPES.

WATER DRAWN IN WHEN OIL AND AIR CONTRACT WHEN COOLED

PREVENTION
1. KEEP BUNGS DRAWN TIGHT. USE WOODEN MALLET TO MAKE SURE.
2. STORE BARRELS INSIDE WHENEVER POSSIBLE.
3. IF STORED OUTSIDE, LAY BARRELS ON THEIR SIDES. IF BARRELS CANNOT BE LAID ON THEIR SIDES, TILT THEM SLIGHTLY AS SHOWN BELOW.

WATER AROUND BUNG MAY BE DRAWN INTO BARREL.

NO WATER AROUND BUNG TO BE DRAWN INTO BARREL.

WRONG   RIGHT

EVAPORATION

Stop evaporation by keeping lids on tight. Loose lids allow fuel to evaporate. A spark or flame can then easily ignite the evaporating fumes.

UNDERGROUND FUEL TANKS

Underground fuel tanks keep fuel cooler, and reduce evaporation. They are safer than aboveground tanks. Underground tanks must continually be checked for water leakage. Remove water from these underground tanks with specially made equipment.

The tanks must also be inspected for fuel leakage into the soil. Absorption into the soil results in a pollutant which requires tank replacement.

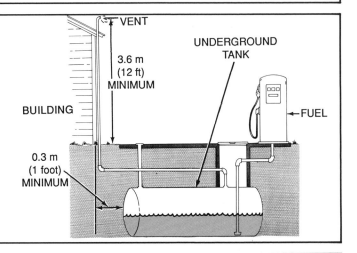

STORING DIESEL FUEL

Evaporation is not a major problem with diesel fuel. But, water and dirt are. Dirt can clog fuel passages. Water can rust parts and make engines run poorly. Water is about the same weight as diesel fuel. It takes about 24 hours for water to settle to the bottom of a fuel tank after the storage tank is refilled. let the water settle before you use diesel fuel.

At least twice a year drain and flush fuel storage tanks.

IMPORTANT: Never store diesel fuel in galvanized (zinc-coated) tanks. Diesel fuel flakes off zinc. The flakes plug up the fuel filters. Use plain steel tanks instead.

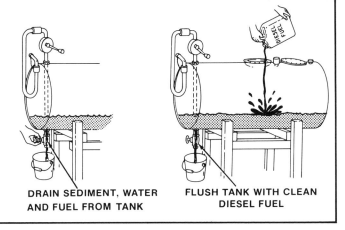

DRAIN SEDIMENT, WATER AND FUEL FROM TANK

FLUSH TANK WITH CLEAN DIESEL FUEL

STORING GASOLINE (PETROL)

Gasoline evaporates easily. Evaporation is a major problem. It robs you of fuel. And worse, the part of the fuel that evaporates first burns best. This means the leftover fuel is hard to burn and makes engines hard to start.

Here is a list of things you can do to reduce evaporation in stored gasoline:

● *You can build a shade over the fuel tank, and paint the tank white or silver.*

● *Install a pressure-vacuum relief valve. It holds most fumes in the tank; only releasing them when pressure gets high. In addition, a pressure-vacuum valve reduces the amount of moist air that enters a tank at night.*

● *NOTE: Do not try to seal a gasoline storage tank completely. If you do, heat from the sun may cause enough pressure inside to burst the tank.*

● *Store gasoline 30 days or less. Longer storage causes chemical changes in the fuel that reduce power.*

● *Drain the water and dirt out the bottom of your gasoline storage tanks two or three times a year.*

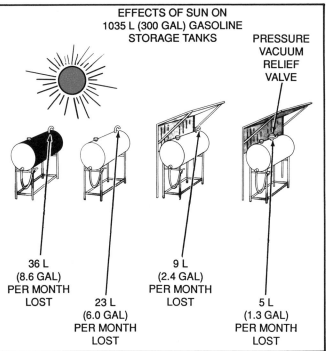

EFFECTS OF SUN ON 1035 L (300 GAL) GASOLINE STORAGE TANKS

PRESSURE VACUUM RELIEF VALVE

36 L (8.6 GAL) PER MONTH LOST

23 L (6.0 GAL) PER MONTH LOST

9 L (2.4 GAL) PER MONTH LOST

5 L (1.3 GAL) PER MONTH LOST

STORING LIQUID PETROLEUM GAS (LPG)

 CAUTION:

● **LPG must be stored in a strong pressure tank that is sealed at all times.**

● **Store LPG away from buildings and other fuel tanks.**

● **LPG fumes are heavier than air. They collect in low spots and closed buildings where they can smother you. A spark or flame can easily ignite them.**

● **Be careful with hoses and connections. Even a small hose leak can lead to a dangerous fire.**

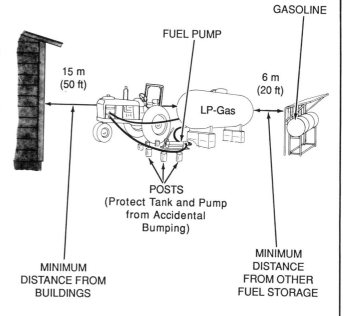

GASOLINE

FUEL PUMP

15 m
(50 ft)

6 m
(20 ft)

LP-Gas

POSTS
(Protect Tank and Pump
from Accidental
Bumping)

MINIMUM
DISTANCE FROM
BUILDINGS

MINIMUM
DISTANCE
FROM OTHER
FUEL STORAGE

 CAUTION: Never use a flame to find a gas leak. The flame could ignite leaking gas. Use soap and water.

If you think a tank, hose, or connection is leaking LPG:

1. **Cover the spot with soap suds and watch for bubbles.**

2. **The bubbles will show you where the leak is.**

3. **You can usually smell the leaking gas, too.**

4. **Have the leak repaired immediately.**

LEAK

SOAP & WATER

DIESEL AND GASOLINE (PETROL) REFUELING

Here are some general ideas for refueling that may help you.

● *Shut off the engine when refueling.*

● *Always use a clean filter on a storage tank outlet to catch dirt and sediment.*

● *Keep the filler cap tight.*

● *Never take fuel from the bottom of a storage tank.*

● *Never use the last 3 or 4 inches (75 to 100 mm) of fuel in the bottom of a storage tank. It is usually dirty or contains water.*

● *Always tip the tank so the deep end is away from the outlet.*

● *Always fill machine fuel tanks directly from the storage tanks or use closed containers to transfer fuel. Do not use a bucket to carry fuel to a machine. Dirt can get in the open bucket and there is a serious danger of fire.*

● *Keep the vent pipes and filter clean.*

● *Fill gasoline and diesel fuel tanks at the end of each work day. If you leave a fuel tank empty overnight, cool, most air will enter, condense into water inside, and run down into the fuel. if the tank is full of fuel there is no room for air.*

⚠ WARNING

A FUEL FIRE CAN BURN YOU.

Tighten cap securely. Do not open if engine is running or hot, near flame, sparks or while smoking. Explosive vapors or fuel may escape. Shut off engine and let it cool before opening cap.

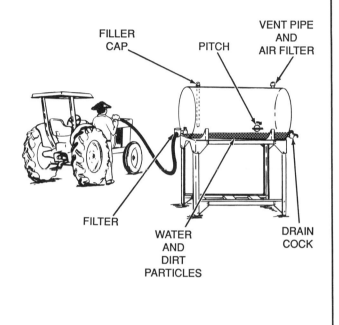

FILLER CAP — PITCH — VENT PIPE AND AIR FILTER

FILTER — WATER AND DIRT PARTICLES — DRAIN COCK

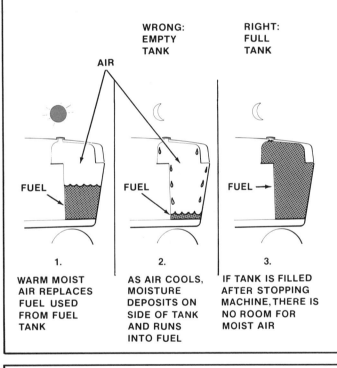

WRONG: EMPTY TANK

RIGHT: FULL TANK

AIR

FUEL

FUEL

FUEL →

1.
WARM MOIST AIR REPLACES FUEL USED FROM FUEL TANK

2.
AS AIR COOLS, MOISTURE DEPOSITS ON SIDE OF TANK AND RUNS INTO FUEL

3.
IF TANK IS FILLED AFTER STOPPING MACHINE, THERE IS NO ROOM FOR MOIST AIR

LPG REFUELING

● *Always wear gloves when refueling. Escaping gas can freeze your skin.*

● *Always use two hoses when refueling. One to carry liquid fuel to the tank being filled, and another to carry fumes back to the storage tank. **Do not vent these fumes into the air.** Escaping fumes catch fire easily.*

● *Never smoke or permit fire or sparks near a machine that is being fueled.*

● *Shut off the engine when fueling.*

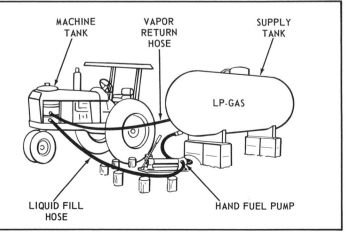

MACHINE TANK — VAPOR RETURN HOSE — SUPPLY TANK

LP-GAS

LIQUID FILL HOSE — HAND FUEL PUMP

DIESEL ENGINE FUEL SYSTEM

Fuel is drawn out of the fuel tank and forced through fuel filters that catch and hold rust, dirt, and some water. Then, an injection pump forces the fuel into the fuel injector on each cylinder. When a piston makes its compression stroke (remember this from chapter 2?) the fuel injector sprays in fuel, which ignites on contact with the hot air in the cylinder, and forces the piston down for the power stroke.

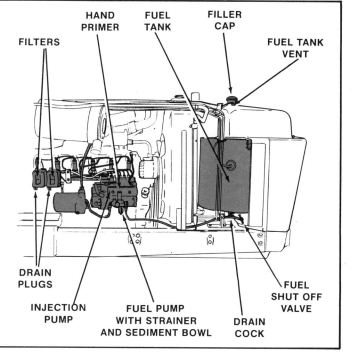

FILTERS

HAND PRIMER

FUEL TANK

FILLER CAP

FUEL TANK VENT

DRAIN PLUGS

INJECTION PUMP

FUEL PUMP WITH STRAINER AND SEDIMENT BOWL

DRAIN COCK

FUEL SHUT OFF VALVE

FUEL TANK

The fuel tank is an important part of a fuel system. It holds fuel in a safe place until the engine needs it.

Fuel tanks for both diesel fuel and gasoline (petrol) systems have a shutoff valve you can close to stop the flow of fuel to the engine when you do maintenance.

Both gasoline and diesel fuel tanks have vents so air can get in to replace the fuel the engine pulls out. A plugged vent can stop fuel flow to the engine. Some tank vents are vented through a pinhole in the cap. Some are vented through a tube. Both should be inspected and cleaned about every 100 hours of operation.

Most diesel and gasoline tanks have a drain valve on the bottom that can be opened to drain out rust, dirt, water and fuel called sediment. Drain sediment out of the tank each time you replace or clean the engine fuel filter.

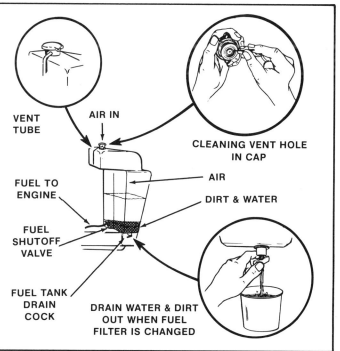

VENT TUBE

AIR IN

CLEANING VENT HOLE IN CAP

FUEL TO ENGINE

AIR

DIRT & WATER

FUEL SHUTOFF VALVE

FUEL TANK DRAIN COCK

DRAIN WATER & DIRT OUT WHEN FUEL FILTER IS CHANGED

DIESEL FUEL PUMP SEDIMENT BOWL

Large pieces of dirt are collected in the bowl before fuel is pumped on through the system.

- *Close the fuel shutoff valve.*

- *If the fuel pump bowl has a drain, open it, and drain out the water and dirt.*

- *Remove the bowl and wash it with a clean cloth and diesel fuel or kerosene. Do not leave dust or lint in the bowl.*

- *Replace the bowl.*

- *Open the drain valve on the bottom of the fuel tank and let 0.5 L (a pint) or so of fuel run out. Sediment in the bottom of the tank should run out with the fuel.*

- *Open the fuel shutoff valve so the engine can get fuel.*

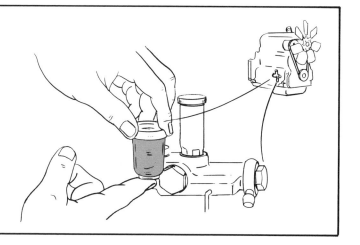

DIESEL FUEL FILTERS

Fuel filters remove damaging rust and dirt from the fuel. They must be cleaned or replaced at least every 500 hours. Some engines have a fuel filter restriction indicator light that comes on when the filter is clogged. It tells you it is time to change or clean the filter.

If the light comes on:

1. **Shut off the fuel flow.**

2. **Drain the filter.**

3. **Remove the filter, housing, and mounting. If the filter is reusable, clean the filter, housing, and mount with clean diesel fuel or kerosene.**

4. **If the filter is not reusable, get a new one.**

5. **Put on the new or cleaned filter and tighten the drain plugs.**

6. **Turn the fuel flow back on.**

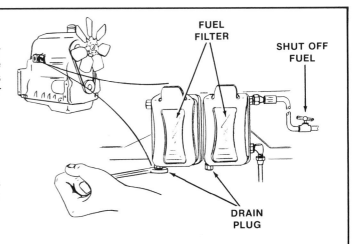

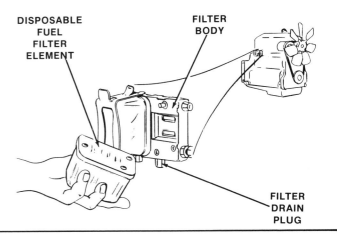

BLEED AIR FROM DIESEL FUEL FILTERS

Air in a diesel fuel system can stop an engine. Air can enter the system if the engine runs out of fuel or when the system is serviced. Air can also get in through a loose or broken line. The air can be removed at the fuel filters:

1. **Loosen the bleed plug.**

2. **Work the hand primer until bubble-free fuel flows out.**

3. **Retighten the bleed plug.**

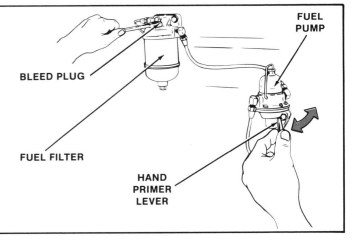

BLEED AIR FROM DIESEL INJECTORS

Air can also be removed at the fuel injector connections. If air is in the system, it will bubble out through a loosened fuel injector connection.

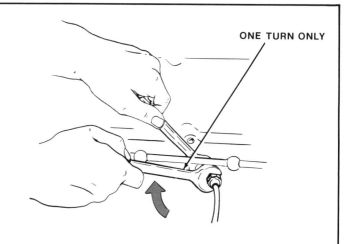

ONE TURN ONLY

 CAUTION: Loosen fittings slowly; diesel fuel, escaping under pressure, can pierce your skin and cause serious injury.

1. Use two wrenches, and carefully loosen the fuel injector lines one turn.

2. Loosen and bleed one injector line at a time.

3. Remember, loosen the injection line nuts one turn only. Fuel will spray out if you loosen them more.

4. Crank the engine with the starter until bubble-free fuel flows out around the loosened fitting. Tighten that line, then loosen and bleed the next one.

5. When you see bubble-free fuel flowing out around a connection, carefully retighten the nut until it is snug and does not leak.

6. Usually you will only need to bleed half of the injection lines.

7. Wipe fuel off the engine.

DIESEL PUMPS AND INJECTORS

Repairing and adjusting diesel injection pumps and injectors requires special training and tools. Always have these repairs made by a diesel mechanic. Do not attempt to adjust, remove, or repair them yourself.

COLD WEATHER STARTING AID

Diesel engines may be difficult to start in cold weather. But, a small amount of starting fluid, sprayed into the cylinders, will usually start them. You can send a shot of starting fluid to the cylinders by pressing the starting fluid can mounted under the dash board or pressing a button. One shot is plenty. Too much starting fluid can blow a hole in a piston when it ignites.

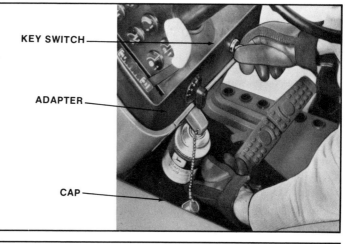

KEY SWITCH

ADAPTER

CAP

● *Do not inject starting fluid unless the starter is cranking the engine.*

●*Replace empty starting fluid cans and keep connections tight.*

● *Never use starting fluid in an engine that is equipped with glow plugs.*

GLOW PLUG STARTING AID

Some diesel engines have small electric heaters mounted in the cylinders to help them start in cold weather. The heaters, called glow plugs, use electricity from the battery to heat the glow plug elements. The hot elements will help to ignite the fuel/air mixture. The plugs are turned on to start the engine and turned off when the engine starts. Never use starting fluid to start an engine equipped with glow plugs.

Only a mechanic should adjust or change glow plugs.

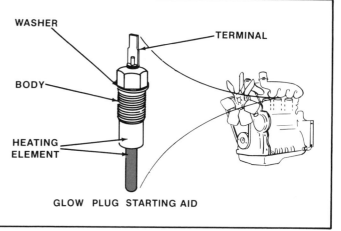

WASHER

TERMINAL

BODY

HEATING ELEMENT

GLOW PLUG STARTING AID

GASOLINE (PETROL) ENGINE FUEL SYSTEM

The main parts of a gasoline (petrol) fuel system are shown in the figure. Gasoline is pumped or flows by gravity from the fuel tank, through a filter, into the carburetor. In the carburetor, fuel is mixed with air. Then, the fuel and air mixture is drawn through the fuel intake manifold into the engine cylinders and burned.

Some engines use a fuel-injection system. Instead of a carburetor there is a fuel-injection pump that sprays fuel into the intake manifold or near each intake port in the cylinder head.

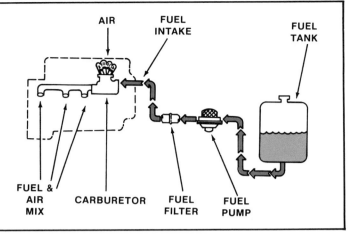

GASOLINE FUEL TANKS

A gasoline fuel tank is much like a diesel fuel tank. It requires the same maintenance (see diesel fuel tank maintenance). It is a closed container with:

- *A vent that must be clean*
- *A fuel shutoff valve that must not leak*
- *A filler cap that must not leak*
- *A drain valve to drain out rust, dirt, and water*

GASOLINE FUEL PUMP

A gasoline engine fuel pump pulls fuel out of the tank, and forces it through a filter, into the carburetor or gasoline (petrol) injection pump. Fuel pumps are so delicate, maintenance should only be done by a mechanic.

GASOLINE ENGINE CARBURETOR

Air is drawn into a gasoline engine carburetor and mixed with gasoline. The mixture is then drawn into the cylinders and burned. A mechanic must adjust the carburetor so it doesn't mix in too much fuel and waste it, or mix in too little fuel and reduce power.

You can help by keeping the air hose tightly connected to the carburetor so air doesn't leak. If the clip isn't tight, replace it. If the hose is damaged, put on a new one. If air leaks out: there will not be enough air in the air-and-fuel mixture; power will drop; and fuel will be wasted.

Check the fuel line connection too. Make sure it doesn't leak. If it does leak:

1. **Shut off the gas flow valve**

2. **Replace the hose or tighten the connection**

3. **Open the gas valve again**

Remember, you can prevent most problems by draining water and sediment out of the carburetor:

1. **Turn off the fuel flow valve.**

2. **Open the carburetor drain plug.**

3. **Drain the gasoline and sediment into a cup.**

4. **Replace the drain plug and tighten it until it's snug.**

5. **Open the fuel flow valve again.**

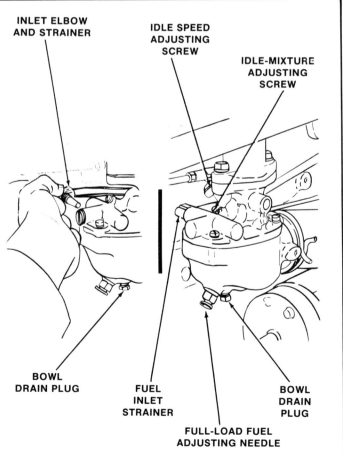

GASOLINE (PETROL) FUEL STRAINER AND SEDIMENT BOWL

The strainer and sediment bowl remove dirt and water from gasoline. Look at the sediment bowl about every 10 hours of engine operation. If you see water or dirt, clean the bowl and strainer as follows:

1. **Close the fuel shutoff valve.**

2. **Loosen the retaining nut and bail.**

3. **Remove nut, bail, and bowl. The strainer mesh and gasket will come off with the bowl.**

4. **Wash the bowl and strainer thoroughly in diesel fuel or kerosene, and wipe dry.**

5. **Inspect the gasket and replace it if it is damaged.**

6. **Reinstall bowl assembly.**

7. **Turn the fuel shutoff valve back on.**

8. **Check for leaks.**

Some gasoline fuel systems have a throw-away filter instead of a strainer and sediment bowl. You can replace a throw-away filter by following the steps above. But, washing isn't required.

Most throw-away filters are located in the fuel line.

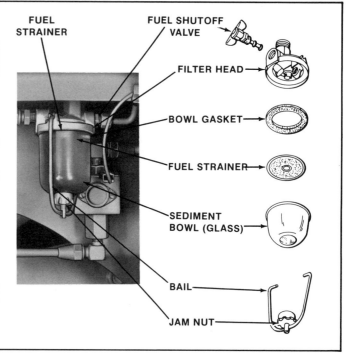

LPG ENGINE FUEL SYSTEM

An LPG fuel system is similar to a gasoline engine fuel system. Liquid fuel is pumped from the pressurized tank, through a strainer into a converter. In the converter, heat turns the liquid into gas. The gas is then mixed with air in a carburetor, and the mixture is drawn into the engine cylinders where it is burned.

As in a gasoline engine, if the carburetor is not adjusted properly by a mechanic, it can burn too much fuel and waste it, or burn too little and reduce power.

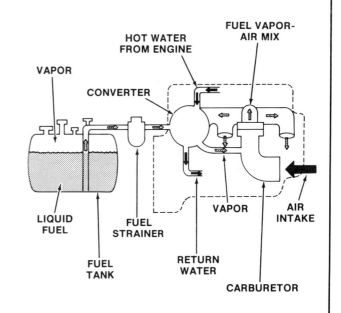

PLUGGED FUEL LINE

Plugged LPG fuel lines are a common problem you can correct. Small particles of dirt can plug lines or keep valves open. Dangerous gas leaks or completely plugged fuel lines can result if you don't remove the plug.

You can see where plugs are in LPG systems because white frost forms where the plug is.

 CAUTION: Never clean or disconnect LPG lines near a flame or spark, or in a closed building. The LPG will blow up if it touches a flame or spark. In the closed building, the LPG can smother you before you know it.

If you see frost on the outside of a line:

1. Shut off the engine and close the vapor valves at the fuel tank.

2. Disconnect the fuel line, and clean it out.

3. Reconnect the line.

If the fuel strainer is plugged, it will get frosty. You can clean it out as follows:

1. Shut off the engine, and close the liquid and vapor lines at the fuel tank.

2. Remove the drain plug from the strainer.

3. Open the vapor valve at the tank slowly to blow the dirt out of the strainer. Then, close the vapor valve.

4. Replace the strainer drain plug, and open the liquid and vapor valves at the tank.

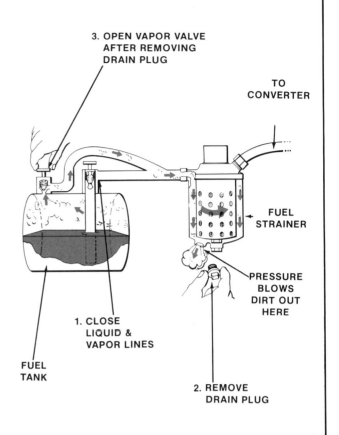

3. OPEN VAPOR VALVE AFTER REMOVING DRAIN PLUG

TO CONVERTER

FUEL STRAINER

PRESSURE BLOWS DIRT OUT HERE

1. CLOSE LIQUID & VAPOR LINES

FUEL TANK

2. REMOVE DRAIN PLUG

SUMMARY

 CAUTION: Fuels are a fire hazard. Use extreme caution!

Fuel is a burnable gas or liquid engines burn to make power. It can be diesel, gasoline (petrol), Liquid Petroleum Gas (LPG), or a mixture of alcohol and petroleum. There are different grades of diesel fuel and gasoline:

● *D-1 and D-2 in the U.S.A. (Winter or summer fuel in the rest of the world.)*

● *Low, regular, and premium gasoline.*

● *Leaded or unleaded gasoline.*

LPG is not graded.

Always burn the fuel recommended for your machine.

Fuel must be stored so it:

● *Is not contaminated by rust, dirt, water, and trash.*

● *Will not evaporate into the air.*

● *Is far away from buildings and equipment.*

When refueling:

● *Keep dirt out of the fuel.*

● *Don't spill fuel on hot engine parts.*

● *Don't smoke or allow fires or sparks near refueling.*

● *Shut the engine off when refueling.*

You can maintain an engine fuel system to keep it running at its best:

● *Drain, clean, and replace fuel strainers and filters.*

● *Keep air hoses and fuel line connections tight.*

● *Replace leaking hoses and gaskets.*

● *Remove dirt plugs from LPG systems.*

DO YOU REMEMBER?

1. What are the two most common fuels in this area?

a.

b.

2. Fuel which starts burning too early in an engine cylinder can cause _____.

3. List three safety rules for fuel storage.

a.

b.

c.

4. When should a gas or diesel tractor be filled with fuel?

5. How often should a diesel fuel filter be replaced?

6. Cold weather diesel engine starting can be improved by using _____.

7. Gasoline engines may have a sediment bowl in the fuel line. True   False

8. To find a leak in LPG lines use _____.

ENGINE OIL SYSTEM
CHAPTER 5

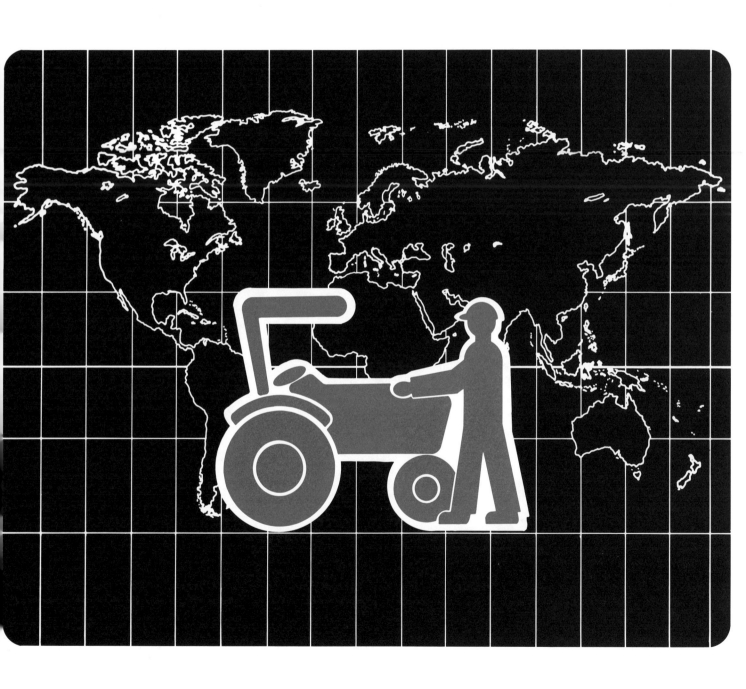

INTRODUCTION

This chapter discusses:

- *What the engine oil system does*
- *The engine oil system parts*
- *Different kinds of oil*
- *Oil additives*
- *Dirty oil*
- *Oil storage*
- *Engine oil system maintenance*

WHAT AN ENGINE OIL SYSTEM DOES

The oil system must do four main jobs for an engine:

1. **Reduce friction between moving parts**
2. **Help cool the engine**
3. **Seal between piston rings and cylinder walls**
4. **Clean engine parts**

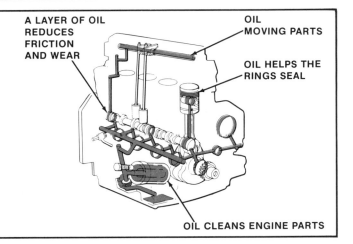

A LAYER OF OIL REDUCES FRICTION AND WEAR

OIL MOVING PARTS

OIL HELPS THE RINGS SEAL

OIL CLEANS ENGINE PARTS

REDUCE FRICTION

Try sliding two dry blocks back and forth on each other. Notice how they scrape. Now, put a film of oil between the blocks and see how much easier they slide.

The oil film keeps the blocks apart so they slide on the oil, not on each other. When engine parts are coated with oil, they too slide easier. The engine oil system keeps a film of oil on all the engine parts to reduce friction.

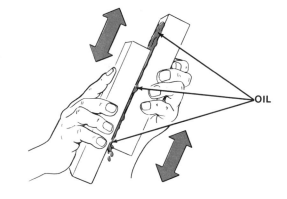

OIL

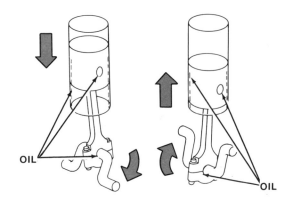

OIL

OIL

HELP COOL THE ENGINE

When an engine is running, oil is pumped up from the crankcase and sprayed on the moving parts. The oil absorbs heat from these parts, runs back down into the crankcase, and radiates the heat out through the crankcase walls. Some machines have oil coolers to help cool the oil.

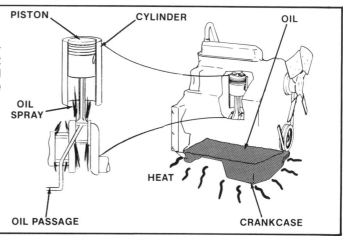

SEAL CYLINDERS

There is a thin space between pistons and cylinder walls so pistons can move. Oil must seal this space to hold pressure in the cylinder during compression and power strokes (remember this from chapter 2?). If the oil seal leaks, much of the engine power is lost because gas leaks between piston rings and cylinder walls.

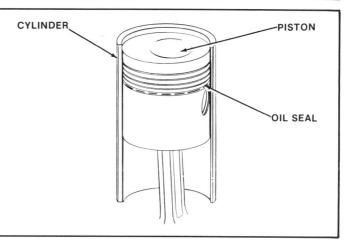

CLEAN THE INSIDE OF THE ENGINE

Oil picks up dirt, small pieces of metal, and some soot as it moves through the engine. Some of these particles are caught and held in the oil filter. But, most settles to the bottom of the crankcase where it can be drained out with the old oil.

If oil is left in an engine too long, it collects so much dirt it can't clean anymore. It also thickens and may plug the filter and oil passages. Also, the cleaning additives in the oil wear out if the oil is not changed regularly.

OLD DIRTY OIL   CLEAN NEW OIL

ENGINE OIL SYSTEM PARTS

Engine oil is stored in the **crankcase.** An **oil pump** draws oil out and pumps it through an **oil filter,** which catches and holds dirt while it lets clean oil through. Then, the filtered oil is forced through **oil passages** to the moving parts of the engine. After the oil does its work of lubricating, sealing, cleaning, and cooling, it flows back to the crankcase. Some engines cool the oil in an **oil cooler** before it goes back into the crankcase. Once in the crankcase, hot oil radiates heat through the metal walls of the case into the air.

An **oil pressure gauge** on the instrument panel tells you if there is enough oil pressure (more on this later).

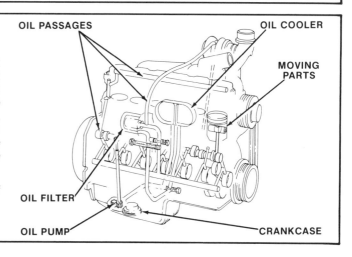

DIFFERENT KINDS OF OIL

Always use the oil recommended in the operator's manual. The oil will be specified by a number like 5, 10, 20, 30, etc., called the viscosity number or weight. Generally, use low numbers, like 5, 10 and 20, in cold weather. The high ones, like 30 and 40, are for hot weather. Your operator's manual will recommend the correct oil for each temperature range.

Oil with more than one number, such as 10W-30, can work in the temperature range for 10 weight through 30 weight. This oil is called a multi-viscosity oil.

Oil is also classified by engine type. The most common classifications are:

- **SF** *oil for gasoline engines*
- **CD** *for diesel engines (also CC)*

There are other classifications of oil, but they should not be used unless specified in your operator's manual. Make sure you use only the recommended kind of oil. Substitutes may not protect your engine properly.

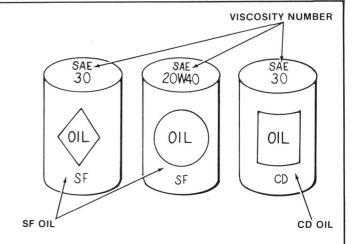

ADDITIVES IN THE OIL MAKE OIL WORK BETTER

Engine oil has chemical additives to help protect engines.

Additives:

- *Reduce rust and corrosion*
- *Reduce oil thickening*
- *Help clean engine parts*

These additives wear out in time. That is one of the reasons for changing the oil regularly.

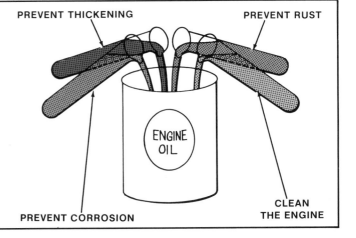

DIRTY OIL

Contaminated oil may contain:

- *Water*
- *Dirt*
- *Unburned fuel*
- *Pieces of metal worn from the engine*

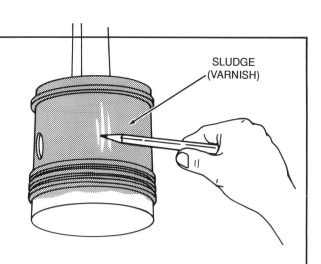

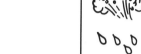

Dirty oil cannot lubricate, clean, or protect your engine. It actually increases engine wear and may fail to seal pistons. Water in the oil rusts engine parts. Thick, contaminated oil sticks to engine parts, plugs oil passages, and forms harmful sludge.

Leaks inside the engine may let engine coolant mix with the oil. If oil looks foamy or milky, it is probably contaminated with coolant. Check for coolant leaks, and change the oil and filter immediately if you see foamy oil on the dipstick. (See chapter 6.)

Dirt and water from dirty oil cans and funnels may get into the oil when you pour oil into the engine.

Condensed water and unburned fuel can also contaminate oil and cause engine damage.

DIRTY OIL MAY BE CAUSED BY:

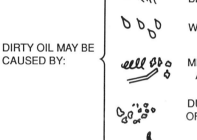

UNBURNED FUEL BLOW-BY

WATER, ACID

METAL BURRS AND CHIPS

DUST, SAND, PIECES OF SEALS AND PAINT

LINT, FIBERS

In a worn engine there is often "blow-by" between piston rings and cylinder walls. This "blow-by" deposits carbon particles in the oil and causes oil thickening.

CONDENSED WATER

When you start an engine, water vapor condenses on the cold metal walls. If the engine continues to run and warm up, the water is changed to steam and vented out. If you don't run an engine long enough to warm up to operating temperature, water will stay inside and rust parts, and causes the oil to become foamy or milky.

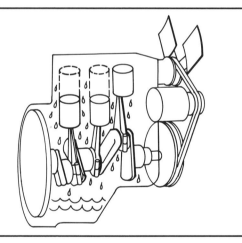

UNBURNED FUEL

In addition to water, unburned fuel collects in the crankcase. The unburned fuel can thin oil until it can't do its job if you don't run an engine long enough to vent it out or burn it. Since fuel contains parafin, the dilution of engine oil with fuel can cause thickening of the oil in cold weather.

This condition is often caused by low engine temperature as a result of a non-functioning thermostat or a removed thermostat.

CRANKCASE VENTILATION

To get rid of the water and unburned fuel in the crankcase, engines have crankcase ventilation systems. There are two kinds:

- *A tube that lets water and fuel escape to the air*

- *A tube that draws crankcase vapor up to the intake manifold and on into the cylinders where it is burned or exhausted with exhaust gases.*

Whichever kind you have, maintain it:

- *Keep the intake and outlet clean and clear*

- *Fix leaks*

- *Follow instructions in your operator's manual*

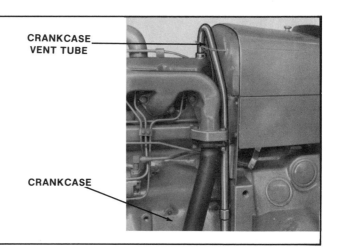

CRANKCASE VENT TUBE

CRANKCASE

STORE OIL PROPERLY

It is best to store oil inside so it is protected from extreme temperatures, dust, rain, etc. Even if you store oil inside, keep the covers on tight.

Oil usually comes in small one quart (1 L) cans, but it may be purchased in large barrels. If you must store oil outside, lay barrels on their sides, so water can't collect on top and seep in around the bungs. If barrels must be stored upright, put a block under one side so rain water can run off. Keep the bungs drawn tight.

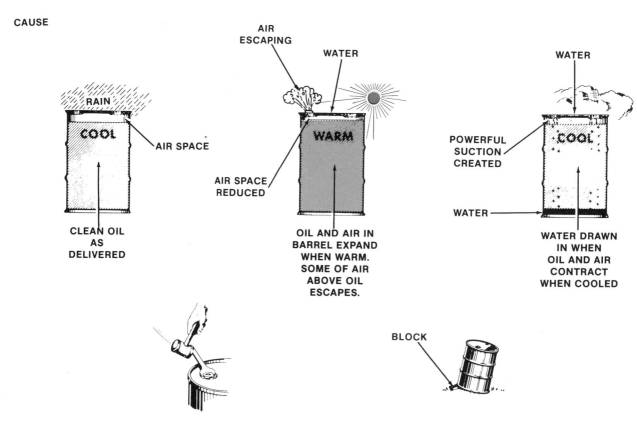

CAUSE

RAIN

COOL

AIR SPACE

CLEAN OIL
AS
DELIVERED

AIR
ESCAPING

WATER

WARM

AIR SPACE
REDUCED

OIL AND AIR IN
BARREL EXPAND
WHEN WARM.
SOME OF AIR
ABOVE OIL
ESCAPES.

WATER

POWERFUL
SUCTION
CREATED

COOL

WATER

WATER DRAWN
IN WHEN
OIL AND AIR
CONTRACT
WHEN COOLED

BLOCK

WATCH THE OIL PRESSURE GAUGE OR INDICATOR LIGHT

Look at the oil pressure gauge often. If the oil pressure reading is abnormal:

1. **Stop the engine.**

2. **Check the oil level in the crankcase, and add more oil if needed. (Be sure you use the correct oil.)**

3. **If pressure remains abnormal, drain engine oil, and change the oil filter. Fill the crankcase with new oil, start the engine, and let it run until it reaches normal operating temperature.**

4. **If the oil pressure is still abnormal, or if the engine makes loud, unusual sounds, turn it off and have it checked by a mechanic.**

Note: The oil pressure gauge does not indicate the amount of oil in the crankcase.

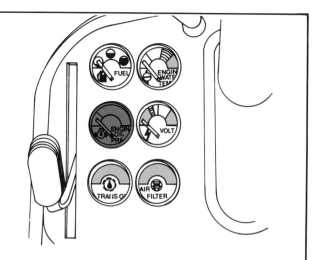

CHECK ENGINE OIL DAILY

Check the engine oil level every day. Most engines have a **dipstick** to measure the oil level; a few have a sight glass. To check the engine oil on an engine with a dipstick:

1. **Stop the engine and let it set for about 5 minutes.**

2. **Remove the dipstick.**

3. **Wipe it off with a clean cloth.**

4. **Reinsert the dipstick; all the way in.**

5. **Remove the dipstick again.**

6. **Look at the oil level on the stick.**

7. **If the oil is below the "Add Oil" mark, pour enough new oil into the engine to raise the level to the "Full" mark. Be sure you use the same kind of oil and wipe dirt off the top of the can and spout before you pour.**

8. **Most engines take about one quart (1 L) of oil to raise the oil level up from the "Add Oil" mark to the "Full" mark on the dipstick.**

9. **While you have the dipstick out, look at the oil carefully. It is all right for it to be dark. After a few hours of engine operation, oil normally turns dark. But, if you feel gritty dirt or pieces of metal in it or if the oil looks milky or foamy, a mechanic should check the engine for damage.**

Some oil loss is normal in engines. Some is burned. Some leaks out around plugs and seals. However, if you have to add oil often, the engine should be examined by a mechanic.

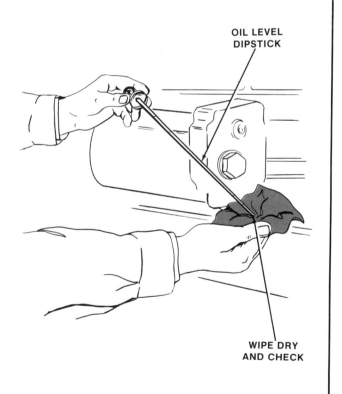

OIL LEVEL
DIPSTICK

WIPE DRY
AND CHECK

HOW TO CHANGE OIL

When oil begins to wear out or accumulate dirt and sludge, it can't protect your engine.

The oil and oil filter must be changed when recommended in your operator's manual. Waiting longer causes wear and may even ruin an engine.

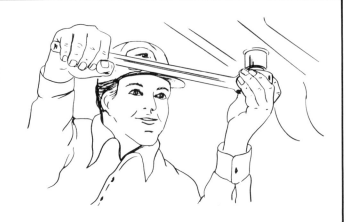

DRAIN THE OLD OIL

Be sure you have:

- *Enough of the recommended oil*
- *The recommended oil filter*
- *Drain plug wrench*
- *Drain pan*
- *A clean cloth*

Before changing oil, run the engine until it reaches normal operating temperature. Running the engine mixes the dirt and sludge in the crankcase with the oil so it can all run out together.

 CAUTION: Do not drain oil that is very hot. Serious burns could result. Wait until it cools.

1. Stop the engine and remove the crankcase drain plug or plugs.

2. Catch all the oil in a bucket. Dispose of the old oil so it doesn't pollute. Some refining companies will buy the used oil.

IMPORTANT: Never operate the engine without oil in the crankcase.

1. Let all the oil drain from the crankcase.

2. If the drain plug is magnetic, clean off pieces of metal carefully. Do not damage the threads.

3. Wipe the drain plugs clean and put them back in the crankcase.

4. Wipe the oil and grime off the crankcase so it can radiate heat.

REMOVE THE OLD OIL FILTER

There are three main kinds of oil filters:

- *Through bolt*
- *Internal*
- *Spin-on*

Before removing the filter:

1. **Wipe dirt and grease from the filter and filter mounting.**

2. **Remove the drain plug and catch the oil in a container.**

3. **Remove the filter.**

4. **Always wipe off the filter mount with a clean cloth. Careful, don't leave lint on the mount.**

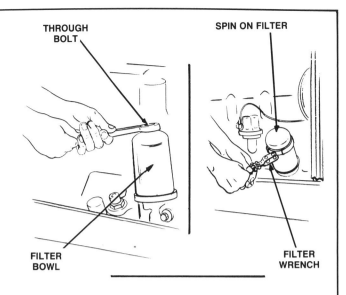

THROUGH BOLT

SPIN ON FILTER

FILTER BOWL

FILTER WRENCH

INTERNAL FILTER

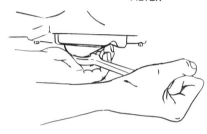

THE FILTER

Most engines have spin-on filters you can discard. Some of these filters have a relief valve built into the filter itself. Make sure you use the right filter.

Some have filter elements that are disposable, but the housings remain with the machine. Wipe these housings clean with a cloth before inserting a new element. Internal filters have a removable cap you must clean and replace.

MOUNT THE NEW FILTER

After the oil is drained, the drain plugs replaced, and you have done your cleanup, mount a new oil filter on the engine.

1. **Wipe off filter base. Leave no lint.**

2. **Use a new filter gasket.**

3. **Spread a thin coat of new oil on the filter gasket.**

4. **Mount the new filter. If it is a spin-on filter, turn it on by hand until the gasket touches the mounting face. Then turn it another half turn until it is snug. Do not overtighten it or you may crush the gasket and cause a leak.**

5. **If you have a through bolt or internal oil filter, insert the correct new filter using a new, lightly oiled gasket. Snug it up, but do not overtighten or the gasket will be crushed and oil will leak out.**

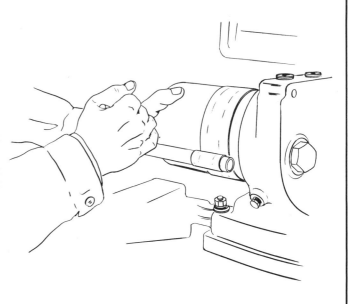

CLEAN OIL CANS AND SPOUTS

Before refilling the crankcase, wipe the dirt off spouts, funnels, and cans. Dirt can be carried into the engine with the oil.

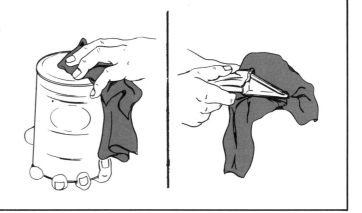

REFILL THE CRANKCASE WITH OIL

After the new oil filter is on and the drain plugs are back in place:

1. **Check your operator's manual to be sure you have the correct amount and kind of oil.**

2. **Open the crankcase filler and pour in the correct amount of oil. Your operator's manual will tell you where to find the filler.**

3. **Do not overfill the crankcase. Overfilling causes oil to foam and leak.**

4. **Let the oil drain down for a few minutes.**

5. **Start the engine and let it run slowly until oil pressure reaches normal.**

6. **Do not run the engine fast until the oil pressure is normal.**

7. **Stop the engine and check around the oil filter and drain plugs for leaks. Tighten the filter and drain plugs, if they leak.**

8. **Check the oil level again. Add oil if the level is too low.**

9. **Wipe spilled oil off the engine.**

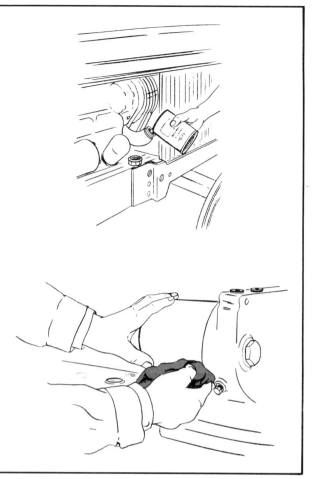

RECORD ENGINE HOURMETER READING

Always write down the date and engine hourmeter reading when you change the oil and oil filter. Then you can figure out when the next service is required. Servicing the engine on time and keeping good records shows good judgment. It saves time and money and reduces the chance of machine failure.

Regular oil and filter changes are an inexpensive investment that will assure long engine life.

SUMMARY

Oil does four jobs for your engine.

- *Reduces friction*
- *Helps cool engines*
- *Seals between pistons and cylinder walls*
- *Cleans the inside of engines*

The engine oil system pumps oil from the crankcase, up through the engine, and back. It also filters the oil and cools it.

There are different kinds of oil for different purposes. Oil is made in different viscosities for hot or cold weather. SF oil is made for gasoline (petrol) engines. CD or CC oil is made for diesel engines.

If in doubt always use CD oils.

Dirty oil containing dirt, water, unburned fuel, carbon, and little pieces of metal will ruin an engine.

Store oil cans and drums so water runs off. Keep the bungs tight so water and dirt cannot get in. Wipe off cans, funnels, and spouts so dirt isn't carried into the engine when you pour in oil.

Check the engine oil every day. Look at the oil on the dipstick to see if it is:

- *Up to the right level*
- *Clean*

If it's low, add oil. If it is dirty, change the oil and filter.

Change oil at the times recommended in the operator's manual.

Changing the engine oil and filter on schedule will help keep your engine running long and well.

DO YOU REMEMBER?

1. List four jobs oil does in an engine.

a.

b.

c.

d.

2. Where is oil stored in an engine?

3. Use oil with a high viscosity number in hot weather. Use oil with a low viscosity number in cold weather.   True   False

4. How can contaminated oil damage an engine?

a.

b.

c.

5. What can cause abnormal engine oil pressure?

a.

b.

6. Why check the oil every day?

7. Why should engine oil be warm before it is drained?

8. Why change oil at the times recommended in the operator's manual?

9. Why keep records of engine service?

ENGINE COOLING SYSTEM
CHAPTER 6

INTRODUCTION

This chapter explains:

- *Why engines must be cooled*
- *The fundamentals of cooling*
- *The parts of a cooling system*
- *Coolants*
- *Cooling system maintenance*

WHY MUST ENGINES BE COOLED?

Engines can only tolerate a certain amount of heat. Unless the heat from burning fuel and friction is controlled, the engine will warp and crack. Some parts will even melt. A cooling system is required to keep the engine temperature in the right range.

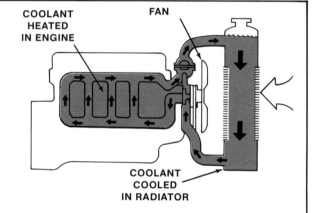

HOW A COOLING SYSTEM WORKS

Engine heat can be controlled with a cooling system. Most cooling systems get rid of extra heat by pumping cooling water through passages in the engine called **the water jacket.** The water absorbs engine heat, just as water in a tea kettle absorbs heat from a hot stove.

After flowing through the water jacket, the water is pumped on to a radiator where it radiates its heat into the air just as a hot stove radiates heat into the air. A fan helps by blowing air through passages in the radiator. After it is cooled, the water is pumped back to the engine for another trip through the water jacket.

Some engines are cooled with air instead of water. A fan blows cooling air over the cooling fins to carry away heat. The most important maintenance for these engines is keeping dirt and trash off so they don't block the air flow and insulate the engine.

COOLANT HEATED IN ENGINE   FAN

COOLANT COOLED IN RADIATOR

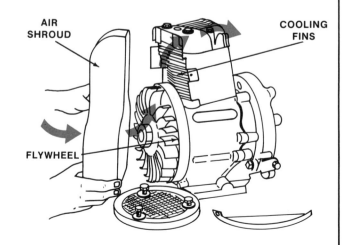

AIR SHROUD

COOLING FINS

FLYWHEEL

COOLING SYSTEM PARTS AND MAINTENANCE

- Liquid coolant
- Radiator cap
- Radiator
- Lower hose
- Upper hose
- Fan and fan belt
- Thermostat
- Engine water jacket
- Water (coolant) pump
- Engine oil cooler (on some engines)
- Coolant filter (on some engines)
- Coolant recovery tank

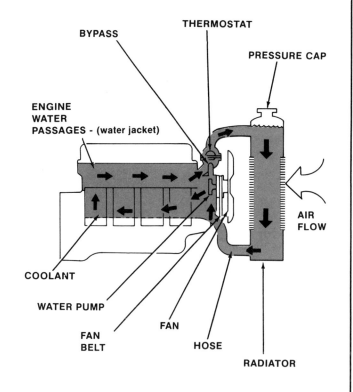

COOLANT

Coolant is clean water with chemical inhibitors mixed in to prevent rust and corrosion. Before you start an engine, while it is still cool, open the radiator cap and look in to see if the coolant is up to the right level; about 1/2 inch (1 cm) from the top. Add clean water if it's low.

Antifreeze is also mixed in to keep the water from freezing in cold weather. Good antifreeze contains rust and corrosion inhibitors, so you don't need to add inhibitors. You can run the water/antifreeze mixture all year long.

Read the directions on any product before you pour it into a radiator. Check the coolant level every day.

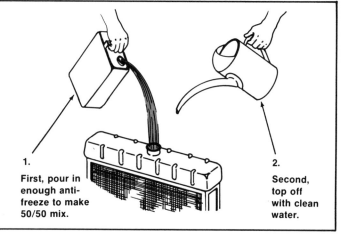

1. First, pour in enough antifreeze to make 50/50 mix.

2. Second, top off with clean water.

RADIATOR CAP

A radiator cap holds pressure in the cooling system. For safety, the cap can open a little to let hot steam out of the overflow tube. It also opens a bit when the engine cools off so air can enter to keep the pressure equal.

About once a year a mechanic should test the radiator cap to see if it holds pressure. You should put a new cap on if the old one won't hold the right pressure or if coolant keeps leaking out.

 CAUTION: Do not remove radiator cap unless engine is cool. Turn cap slowly to release all pressure before removing cap.

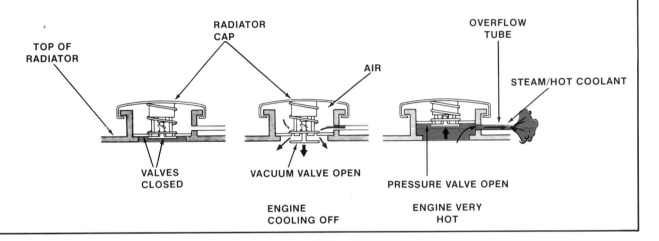

RADIATOR

Air must be able to blow through the radiator. Brush the dirt and chaff off the radiator screen every day. If the screen gets clogged, the fan can't pull air through to cool the coolant, and the engine will overheat.

Once or twice a year — more often if dirt builds up and causes overheating — clean the dirt off the radiator fins. A water hose will usually do a fine job if you squirt the water from the inside out. Compressed air will also work. Be careful. Don't accidentally bend the fins while cleaning. Bent fins reduce air flow even more than dirt.

COOLANT RECOVERY TANK

Some engine cooling systems have a coolant recovery tank. Coolant level in the tank should be 25 to 50 mm (1 to 2 in.). If coolant level is low, fill it through filler.

HOSES AND CLAMPS

Look at the hoses and clamps often. If you see a leak by a clamp, tighten the clamp until the leak stops. If it won't stop, replace the hose. Replace hoses when they get soft, weak, or cracked. When you replace a hose, use sealing compound and tighten clamps so they don't leak.

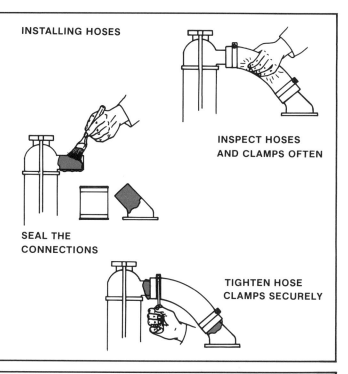

INSTALLING HOSES

INSPECT HOSES AND CLAMPS OFTEN

SEAL THE CONNECTIONS

TIGHTEN HOSE CLAMPS SECURELY

FAN BELT

Don't operate with the fan belt too tight or too loose. If a belt is too tight, it can damage bearings and wear itself out fast. If it's too loose, it will slip, wear out fast, and turn the fan and water pump too slowly to cool the engine. To adjust:

1. Loosen the bolts holding the alternator, or generator.

2. Move the entire alternator or generator to adjust.

3. When the belt is adjusted correctly, you should be able to wiggle the belt about as far as it is between the knuckles of your hand. There are precise fan belt tension specifications listed in operator's manuals.

4. Retighten the alternator or generator bolts after adjustment.

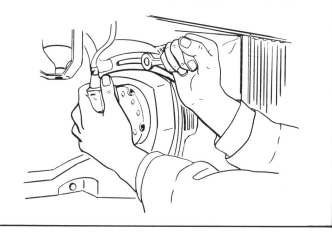

THERMOSTAT

Engines don't need to pump coolant through the radiator when the engine is cool. So, a control valve, called a thermostat, is mounted in the cooling system to let coolant into the radiator when the engine is hot or keep it in the water jacket when the engine is cool. When the coolant gets hot, the thermostat opens and coolant flows through the top hose into the radiator. This keeps an even temperature in the cylinder block for proper combustion.

An engine that doesn't operate at the normal temperature may have a broken thermostat. A mechanic can check and put in a new thermostat if necessary.

IMPORTANT: Never run an engine without a thermostat.

WATER JACKET

Leaks in the water jacket can stop an engine. A leak can be seen easily if the coolant leaks out on the ground. But, if it leaks inside the engine, the only way to tell is if you see bubbles in the coolant or the engine oil looks milky. More on this under **HOW TO FIND LEAKS.**

THE WATER PUMP

When a water pump breaks or wears out it cannot pump coolant through the engine and radiator. The water pump will squeal, and leak coolant. Also, the engine may overheat. A mechanic, with the proper training, should replace the water pump.

ENGINE OIL COOLER

The oil temperature will go up fast if the oil cooler breaks. Oil will thicken and lose its ability to lubricate. If the temperature goes up, check the following:

- *coolant level*
- *leaks*
- *plugged hoses*
- *fan belt tension*
- *thermostat*

If these are all right, the problem is probably in the oil cooler. A mechanic will have to repair or replace it.

HOW TO DRAIN AND FLUSH THE COOLING SYSTEM

Antifreeze and cooling system inhibitors wear out. So, at least once a year, drain and flush the cooling system and refill it with a fresh coolant mixture.

 CAUTION: Be careful. Don't burn yourself. Wear gloves or use a rag to slowly open the drain valves and remove the radiator cap.

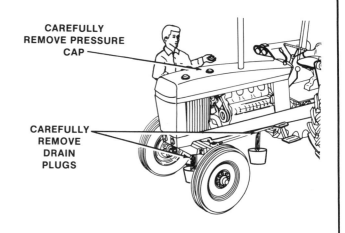

CAREFULLY REMOVE PRESSURE CAP

CAREFULLY REMOVE DRAIN PLUGS

1. Start the engine and let it warm up.

2. Stop the engine and open the radiator cap one stop. Let the steam escape. Then, take it off. Let the system drain completely. Don't forget to drain coolant from the oil cooler, if your machine has one. Most engines have a drain valve on the bottom of the radiator and one or more drains on the engine block. Open them all. See your operator's manual for their exact locations.

3. Close the drain valves.

4. Flush the system by refilling it with a mixture of clean water and a cooling system cleaner recommended in your operator's manual. Carefully follow directions on the cleaner container. Start and run the engine until it is up to normal operating temperature again.

CLEANER

5. Open all the drain valves and pressure cap again, and completely drain the cleaner out of the engine. Then flush the cooling system with fresh water.

6. Once the engine cooling system is flushed clean, close the drain valves and pour antifreeze, rust inhibitors in warm climates, into the radiator until it's half full. Then, fill it up to about 1 cm (1/2 inch) of the top with clean water.

7. Put the radiator cap back on.

8. Start the engine and let it run until it is completely warmed up to mix the water and antifreeze.

9. You may need to add a little water to bring the coolant up to the correct level.

IMPORTANT: Never run an engine without coolant; even for a few minutes. It will overheat.

ANTIFREEZE

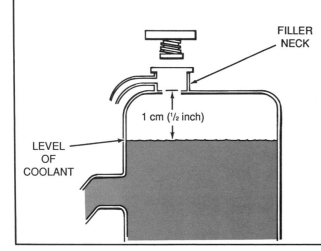

FILLER NECK

1 cm (1/2 inch)

LEVEL OF COOLANT

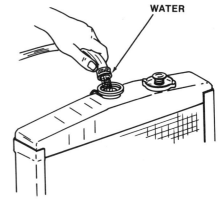

WATER

HOW TO FIND LEAKS

Look at all the hose connections, drain valves, and the radiator cap. Tighten leaking connections or valves. You will need to replace the radiator cap if it leaks.

A crack in the water jacket or a damaged gasket can let coolant leak out or engine exhaust gas leak into the water jacket. Exhaust gas destroys inhibitors in the coolant and forms acid which corrodes, rusts, and clogs the cooling system.

As stated earlier, coolant leaking out of the water jacket is easy to see. It is harder to find an exhaust gas leak.

To find exhaust gas in the water jacket:

1. Start the engine and leave it running while you look.

2. Carefully remove the radiator cap before the engine heats up. Look for bubbles in the coolant. Bubbles or an oil film on the coolant mean exhaust gas may be leaking into the cooling system.

3. Look for bubbles before the engine becomes hot enough for coolant to boil or you could mistake steam bubbles for exhaust gas bubbles.

4. If you see bubbles, have a mechanic repair the system.

5. Pull out the oil dipstick.

6. Look at the appearance of the oil. If it is milky you are probably getting coolant into the oil.

7. Have a mechanic check the system.

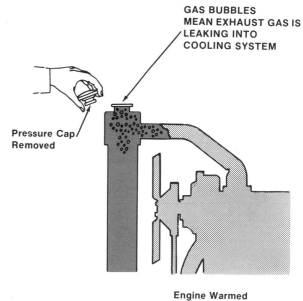

GAS BUBBLES MEAN EXHAUST GAS IS LEAKING INTO COOLING SYSTEM

Pressure Cap Removed

Engine Warmed Up and Under Load

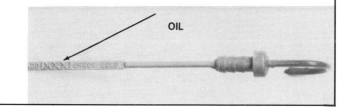

OIL

HOW TO DEAL WITH AN OVERHEATED ENGINE

⚠️ **CAUTION: Stay clear of hot engine parts, steam, and hot coolant. You could get seriously burned. Protect yourself properly.**

Sometimes an engine will get too hot. The temperature gauge will tell you. Sometimes steam and smells will come from the engine.

Do not stop an overheated engine and pour cold water into the radiator. Let the engine idle. Cold water on a hot engine can crack the engine just as ice water can crack a hot glass. Let the engine idle so the temperature can drop as low as possible. Then, carefully:

1. Use a cloth or glove to protect your hand, and carefully turn the cap to the first stop. Stand back and let steam and hot coolant escape.

2. Take off the cap.

3. With the engine still running, add water slowly. Running the engine lets the water mix with the hot coolant gradually so the engine doesn't crack.

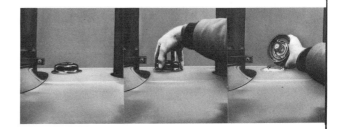

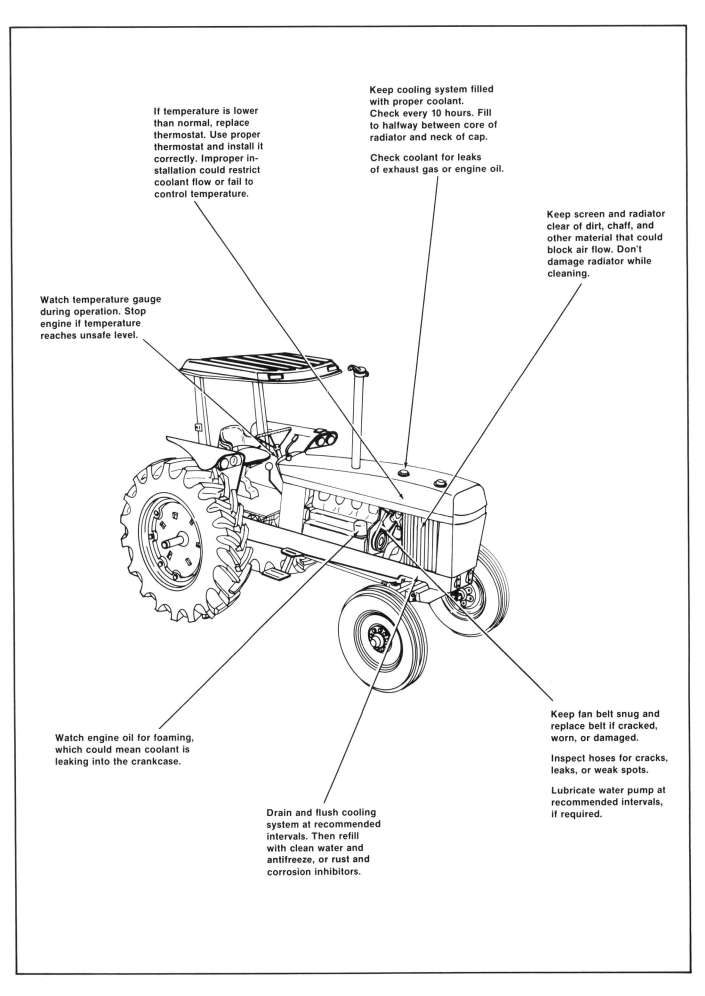

If temperature is lower than normal, replace thermostat. Use proper thermostat and install it correctly. Improper installation could restrict coolant flow or fail to control temperature.

Keep cooling system filled with proper coolant. Check every 10 hours. Fill to halfway between core of radiator and neck of cap.

Check coolant for leaks of exhaust gas or engine oil.

Keep screen and radiator clear of dirt, chaff, and other material that could block air flow. Don't damage radiator while cleaning.

Watch temperature gauge during operation. Stop engine if temperature reaches unsafe level.

Watch engine oil for foaming, which could mean coolant is leaking into the crankcase.

Drain and flush cooling system at recommended intervals. Then refill with clean water and antifreeze, or rust and corrosion inhibitors.

Keep fan belt snug and replace belt if cracked, worn, or damaged.

Inspect hoses for cracks, leaks, or weak spots.

Lubricate water pump at recommended intervals, if required.

SUMMARY

Liquid-cooled engines pump cooling liquid through the water jacket. The coolant absorbs heat in the engine and radiates it into the air through the radiator.

The main maintenance tasks you can do to keep the engine cooling system working are shown in the figure. Pay special attention to the coolant level, fan belt tension, dirt on the radiator, and leaks.

Air-cooled engines blow air over cooling fins to cool off. They must be kept extremely clean so dirt, grease, and trash don't block the air flow and insulate the engine.

DO YOU REMEMBER?

1. What can happen if engine heat is not controlled?

a.

b.

2. An engine can cool itself more than necessary. What part keeps the engine cooling system adjusted to the correct temperature so it only cools itself the correct amount?

3. Why not use plain water in a cooling system?

4. Antifreeze lasts for the life of a machine.   True   False

5. How can you tell if an engine is too hot?

a.

b.

c.

6. How do you maintain an air cooled engine's cooling system?

ELECTRICAL SYSTEM
CHAPTER 7

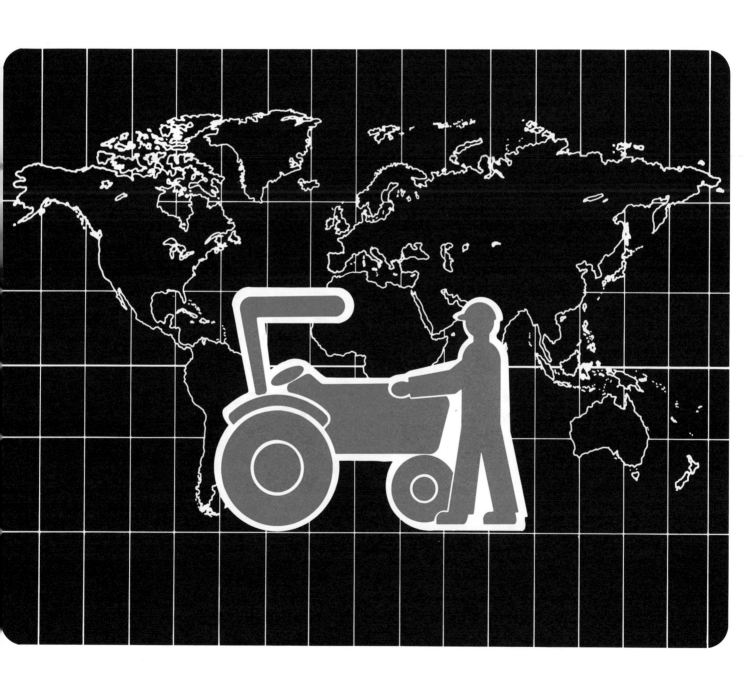

INTRODUCTION

This chapter will explain:

- *What an electrical system is*
- *What an electrical system does*
- *Electrical system parts*
- *Maintenance*

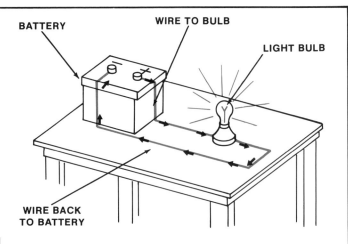

WHAT IS AN ELECTRICAL SYSTEM?

To understand an electrical system, you must first understand an electrical circuit. An electrical circuit is a path that electric current follows. The figure shows the path electricity takes from a battery, through a wire to a light bulb, and back again to the battery. As long as electricity flows through the circuit, the light bulb glows. But, if the circuit is broken, the bulb will go out. The electrical system in your machine works the same way, except for two factors:

- *It has many circuits, because there are many lights, electric motors, and other devices which use electricity.*

- *There are no wires back to the battery from the devices. Instead, the devices are connected to the metal frame of the machine, and electricity flows back to the battery through the frame.*

- *Sometimes a direct ground wire is used to reduce interference (noise) in electronic equipment.*

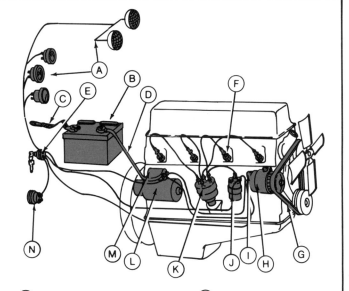

WHAT DOES AN ENGINE ELECTRICAL SYSTEM DO?

Use this drawing for reference as you learn about the electrical system.

There are four circuits in an electrical system. They are named for what they do:

- **Starting circuit** *to start the engine*

- **Charging circuit** *to make electricity and store it in the battery*

- **Ignition circuit** *to make a spark to ignite fuel in spark ignition engines*

- **Accessory circuits** *to power lights, instruments, and accessories*

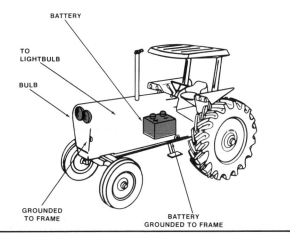

(A)	INSTRUMENTS, LIGHTS, AND ACCESSORIES	(H)	ALTERNATOR
(B)	BATTERY	(I)	VOLTAGE REGULATOR
(C)	GROUND CABLE TO FRAME	(J)	COIL
(D)	POWER CABLE	(K)	DISTRIBUTOR
(E)	STARTER SWITCH	(L)	STARTER
(F)	SPARK PLUG	(M)	STARTER SOLENOID
(G)	FAN BELT	(N)	VOLTMETER (OR AMMETER)

STARTING CIRCUIT

The **starting circuit** connects the battery to a starter switch and an electric starter motor. When the starter switch is turned on, electricity flows to the starter motor solenoid switch.

The solenoid switch directs electricity to the starter motor. The starter motor cranks the engine by turning the flywheel.

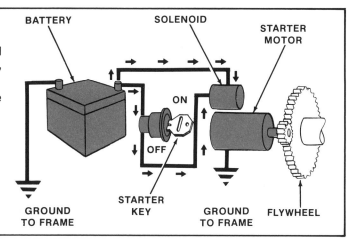

CHARGING CIRCUIT

The charging circuit charges the battery with electricity and provides electricity to power electrical devices while the engine is running. The circuit has:

- An alternator, or a generator, to generate electricity

- A voltage regulator, to maintain an even flow of electricity

- A voltmeter, or ammeter, to show if the alternator or generator is working

- A starter switch (also called ignition switch), to start and stop the flow of electricity to the starter and sometimes to other accessories

- A storage battery to store electrical energy

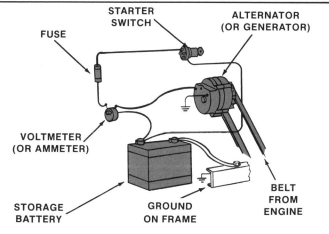

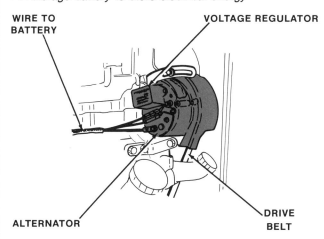

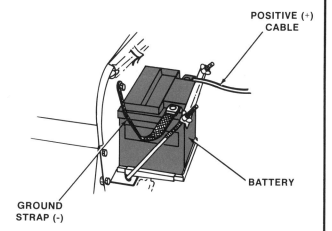

NOTE: Some electrical systems have a positive (+) ground and a negative (−) power cable connection. The usual circuit in modern machines has a negative (−) ground and a positive (+) power cable. This is called polarity. Check your operator's manual for the correct circuit design and polarity. If you make a wrong connection, you may cause serious damage to the electrical system.

IGNITION CIRCUIT (SPARK IGNITION ENGINES)

Engines which burn gasoline (petrol), LPG, or natural gas require an electric spark in the cylinder to ignite fuel. The ignition circuit uses a spark plug in each cylinder to make that spark.

A coil and condenser are used to increase electrical voltage from the battery so a big spark can be made by the spark plugs. A distributor sends electricity to each spark plug at the correct time for the power stroke. Remember this from Chapter 2?

On many small engines, such as those in chain saws and lawn mowers, a magneto instead of an alternator or generator makes electricity for the spark plugs.

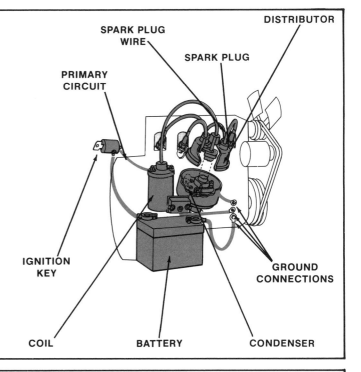

ACCESSORY CIRCUITS

Electricity is also used by lights, instruments, cigarette lighters, diesel engine glow plugs, and controls. Electricity flows from the battery, through the instrument and accessory circuit wiring harness, to these devices.

NOTE: Broken wires and loose connections stop electricity flow. Keep wires away from moving parts, which could fray or cut them.

Signal gauges and lights alert you to problems such as low oil pressure, engine overheating, and alternator problems. Watch the gauges for signs of problems. Signal lights also alert you. These lights come on when the starter switch is turned on. They should go off within a few seconds after the engine starts if everything is working properly.

If gauges or signal lights do not work when the starter switch is turned on, you probably need to tighten a connection or replace a burned out light bulb. You may also have a blown fuse. These signal lights and gauges are important. If they don't work, a problem in the machine could go undetected until it causes serious damage.

(A) TAIL LIGHT

(B) FLASHER WARNING LAMP

(C) HEADLIGHTS

(D) WIRING HARNESS

(E) LIGHT SWITCH

(F) DASH LAMP

(G) ELECTRICITY FROM BATTERY

(H) OIL PRESSURE INDICATOR LAMP

(I) ALTERNATOR INDICATOR LAMP

(J) CIRCUIT BREAKERS

(K) OUTLET SOCKET

(L) FUSES

Gauges and signal lights on the instrument panel.

FUSES AND CIRCUIT BREAKERS PROTECT THE CIRCUITS

The circuits must be protected from overloads and short circuits, also called "shorts." Short circuits and overloads happen when wires wear thin or break, or when electrical devices break or draw too much current. Burned wires and destroyed equipment may result.

Fuses and circuit breakers protect the circuits from shorts and overloads. Circuit breakers are little switches that open when an overload of electricity tries to go through. They can be reset and left in service. Some reset themselves. If a fuse burns out it cannot be reset. The bar inside the fuse melts. The entire fuse must be replaced with another fuse of exactly the same size and number. The wrong fuse could let too much electricity through, damage wiring and equipment, and result in a fire.

If fuses blow or circuit breakers open frequently, have a mechanic find the cause and fix it. The common causes are:

- *Broken wires and bare wires that touch "ground"*
- *Loose connections where wires attach to devices*
- *Poor connection between electrical devices and the machine frame*
- *Water inside light fixtures*
- *Overloaded circuits caused by extra lights and other add-on electrical devices that use electricity*
- *Wrong fuses in the circuit*
- *Fuses too close to the hot engine*

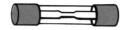

QUICK-BLOWING FUSE
- **If blown from overload - glass will be clear.**
- **If blown from short circuit - glass will be dark.**

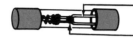

SLOW-BLOWING FUSE
- **If blown from overload - fuse will be broken at solder here.**
- **If blown from short circuit - wires will be burned out here.**

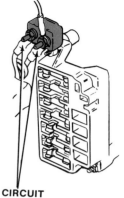

CIRCUIT BREAKER

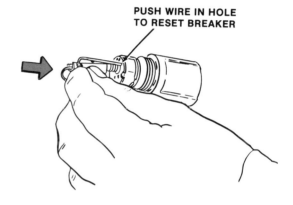

PUSH WIRE IN HOLE TO RESET BREAKER

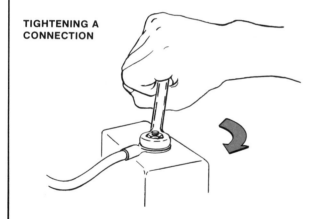

TIGHTENING A CONNECTION

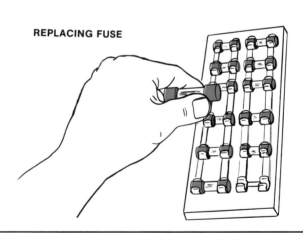

REPLACING FUSE

BATTERY MAINTENANCE

Batteries need regular maintenance. Neglected batteries fail before they should, cost extra money to replace, and delay field work.

WHAT CAN GO WRONG?

Here are the common things that can go wrong with batteries:

● **Overcharging** *evaporates water from the battery fluid. If the battery fluid is often low, a mechanic should check the charging circuit for overcharging.*

● **Undercharging** *weakens battery power. The voltmeter or indicator light on your instrument panel will tell you if the charging circuit is undercharging. The charging circuit should be examined by a mechanic if you notice undercharging.*

Undercharging can be caused by:

● *Frequently starting and stopping the engine without letting it run long enough to fully charge the battery*

● *A loose alternator drive belt*

● *A broken alternator or voltage regulator*

● *Loose connections between the alternator and battery*

● *Hot weather, because batteries discharge faster*

If you add electrical devices, such as extra lights, you may need a more powerful alternator to keep the battery charged.

Other battery problems can be caused by:

● *Low battery fluid*

● *Corrosion on terminals*

● *Cracks in the battery case*

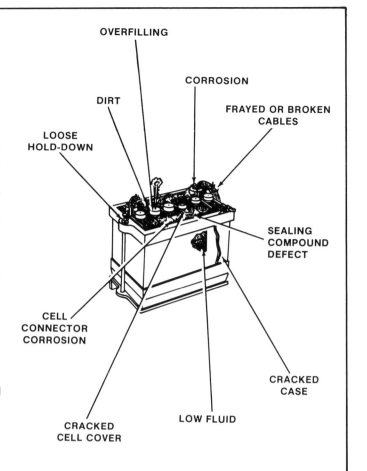

OVERFILLING

CORROSION

DIRT

FRAYED OR BROKEN CABLES

LOOSE HOLD-DOWN

SEALING COMPOUND DEFECT

CELL CONNECTOR CORROSION

CRACKED CASE

CRACKED CELL COVER

LOW FLUID

PREVENT ACID BURNS

 CAUTION: Sulfuric acid in battery electrolyte is poisonous. It is strong enough to burn skin, eat holes in clothing, and cause blindness if splashed into eyes.

Avoid the hazard by:
1. Filling batteries in a well-ventilated area.
2. Wearing eye protection and rubber gloves.
3. Avoiding breathing fumes when electrolyte is added.
4. Avoiding spilling or dripping electrolyte.
5. Use proper jump start procedure.

If you spill acid on yourself:
1. Flush your skin with water.
2. Apply baking soda or lime to help neutralize the acid.
3. Flush your eyes with water for 10-15 minutes. Get medical attention immediately.

If acid is swallowed:
1. Drink large amounts of water or milk.
2. Then drink milk of magnesia, beaten eggs, or vegetable oil.
3. Get medical attention immediately.

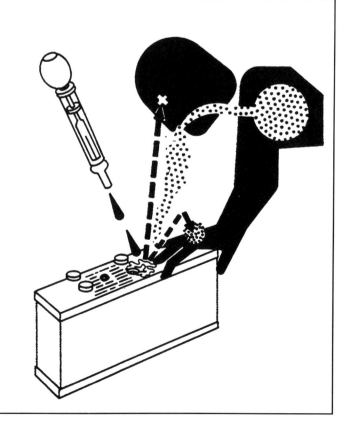

BATTERY SAFETY

Keep all sparks and flames away from batteries. A flame or spark can explode the gas given off when batteries are charging. Battery fluid (acid) can cause serious burns if spilled on your skin. It will also destroy clothing and paint. Please follow these safety steps whenever you work with a battery:

● *If battery fluid spills on your clothing, remove the clothing immediately.*

● *If battery fluid touches your skin, wash with clean, running water for 10 to 15 minutes.*

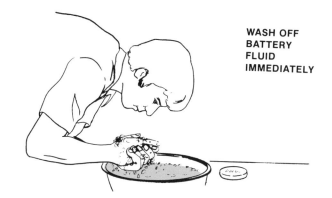

WASH OFF BATTERY FLUID IMMEDIATELY

BE CAREFUL CARRYING BATTERIES.

● *If battery fluid splashes into anyone's eyes, force their eyelids open and flood them with clean water for 10 to 15 minutes. Then see a doctor.*

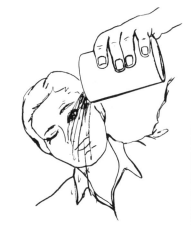

FLUSH EYES

● *Clean up spilled battery fluid with a mixture of 500 grams (1 pound) of baking soda in 4 liters (1 gallon) of water, or a liter (quart) of household ammonia in 8 liters (2 gallons) of water. Then flush it with clean water.*

WATER   BAKING SODA

- *Always disconnect the battery ground cable, the one going to the frame, before you work on any part of the electrical system. This will prevent short circuits or an accidental engine startup.*

- *Always disconnect the ground cable **first** when you take a battery out of a machine, and attach the ground cable **last** when you put a battery in.*

- *Disconnect the battery cables before you recharge a battery so you don't damage the alternator.*

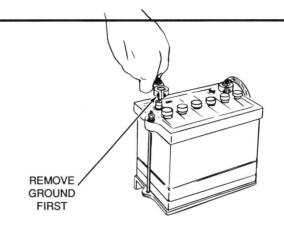

REMOVE GROUND FIRST

- *Never lay a tool or other piece of metal across the battery terminals to check for a spark. It won't tell you how good the spark is, and the spark could explode the battery gas.*

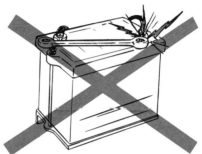

BATTERY CABLES TO STARTER

SINGLE BATTERY

- *Be sure the battery cables are connected to the correct terminals. If you connect the ground strap and power cable to the wrong terminals, you will damage the alternator, and the battery will not charge properly.*

- *Check power cable for insulation wear or damage. A short circuit will discharge the battery and could cause a fire or an explosion.*

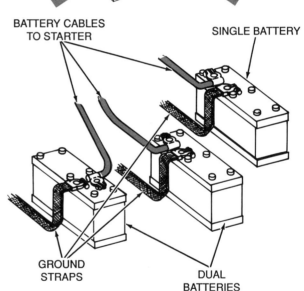

GROUND STRAPS

DUAL BATTERIES

PREVENT MACHINE RUNAWAY

Avoid possible injury or death from machinery runaway.

Do not start engine by shorting across starter terminals. Machine will start in gear if normal circuitry is bypassed.

NEVER start engine while standing on ground. Start engine only from operator's seat, with transmission in neutral or park.

BATTERY FLUID

Every 50 operating hours, or once a week, remove the vent caps and look at the battery fluid. See if it comes up to the bottom of the vents. If it's low, add distilled water until the fluid covers the plates and comes up to the bottom of the vent cap. Don't overfill the battery or the fluid will bubble out and corrode the terminals.

If distilled water is not available, put in clean rain water. But, pour it through a clean cloth first to filter out dirt. Dirty water and ground water contain minerals that damage batteries.

Never add water to a battery in freezing weather unless you run the machine at least a half hour. Otherwise the water won't mix with the battery fluid and will freeze and crack the battery.

⚠️ CAUTION: If you overfill a battery, fluid will spray out through the vent caps when the battery is charging. Electricity can then be conducted out of the battery through this fluid. The spilled fluid will also corrode the terminals and metal around the battery.

Maintenance-free batteries are sealed so you can't add water. However, you need to clean them regularly and protect them from corrosion, cracking, overcharging, and undercharging.

Keep batteries charged in cold weather. In a discharged battery, the electrolyte turns to water. It will freeze and crack the battery.

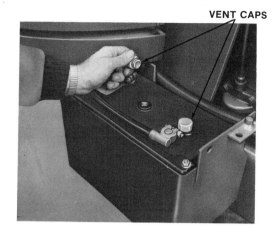

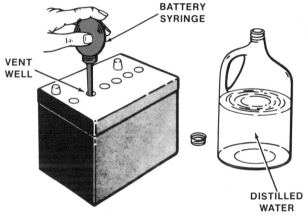

HOW TO CLEAN A BATTERY

About twice a year, or whenever the battery gets dirty, wipe the battery clean with a damp cloth. If there is a white substance on the battery terminals and holddown clamps, clean the battery as follows:

1. Remove the ground cable. Don't pound on the terminal or twist off the clamp. You could crack the battery case or loosen the posts inside.

2. Take off the second battery cable and lift out the battery. Don't tip it or hold it against your clothing. You could be burned by leaking fluid. Use a battery lifting handle if you have one.

3. Wash the cable clamps in a mixture of baking soda and water 57 g of soda in 1 liter of water (2 oz. soda in 1 qt. water). Use a scrub brush to remove the dirt and corrosion and then dry the clamps. Then, polish the inside of each clamp with a round wire brush or a small piece of sandpaper rolled around your finger (rough side out).

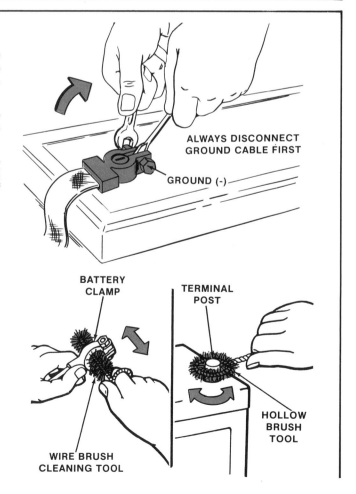

(continued on next page)

4. Wash off the battery case and battery holder with some baking soda solution and a scrub brush.

NOTE: Be sure the vent caps are closed tightly before you wash the battery case. Don't get baking soda in the battery. Soda will weaken the battery fluid.

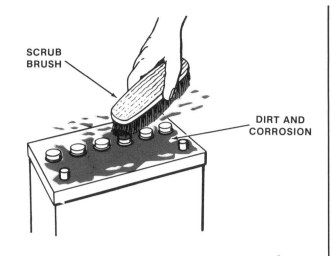

SCRUB BRUSH

DIRT AND CORROSION

BAKING SODA AND WATER SOLUTION

5. Flush the battery with clean water carefully. Don't get any water down the vent caps.

6. Dry the battery with a clean cloth and look closely for cracks.

7. Put the battery back in the machine.

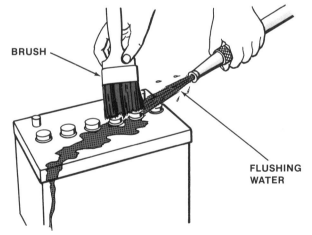

BRUSH

FLUSHING WATER

8. Smear some grease or petroleum jelly (vaseline) on the terminal posts and cable clamps to protect them from corrosion.

9. Reconnect the cables. Connect the ungrounded cable first and be sure you connect positive (+) to positive (+) and negative (−) to negative (−).

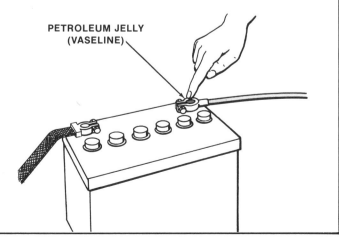

PETROLEUM JELLY (VASELINE)

HOW TO RECHARGE A BATTERY

If your battery doesn't have enough power to start the engine, the problem is probably one of the following:

● *The alternator drive belt may be loose. If it is, tighten the belt as shown in Chapter 6.*

● *Lights or other electrical equipment may have accidentally been left on while the engine was shut off and drained the battery. If so, recharge the battery and remember to turn off the switches next time.*

● *You may have cranked the starter too long and drained the battery. Again, you should recharge the battery.*

● *If the engine hasn't been run for several months, the battery may have lost its charge. Recharge batteries every 30 days during storage.*

● *The battery, alternator, or voltage regulator may have been damaged and require replacement.*

Unless a discharged battery is worn out or broken, it can be recharged with a battery charger, as follows:

1. **Disconnect the battery ground cable first.**

2. **Then, disconnect the ungrounded cable. Use a wrench and puller. Don't pound on the battery post.**

3. **If the battery charger has a switch for 12 or 6 volts, switch to the same voltage as the battery; usually 12 volt.**

4. **Remove the battery vent caps, and add distilled water if fluid is low. Cover the vent openings with a cloth to keep dirt out and to keep battery fluid from spraying out during charging.**

5. **Connect the cables from the charger. If you are charging a battery on a negative ground machine, connect the positive (+) cable on the charger with the positive (+) post on the battery first. Connect the negative (−) cable to the negative (−) post second. Always follow the battery charger and machine operator's manual for specific hook-up.**

6. **Turn on charger by plugging it in and pressing the "charge" button if it has one.**

7. **Slow charge the battery for 12 to 24 hours if it is completely discharged.**

8. **If you know the battery has only been discharged a day or two, "fast charge" it for several minutes. One way to tell if the battery is charged is to look at the voltmeter on the charger when you first turn it on. When the voltmeter has dropped to half that original charge rate, the battery is usually charged.**

9. **When the battery is charged, turn off and unplug the charger, and remove the charger cables.**

10. **Reconnect the battery cables; ungrounded cable first, grounded cable second. Be sure you put the cables back on the same terminals they came off of. If the cables are reversed, the current flows backwards through the circuit and can damage the alternator.**

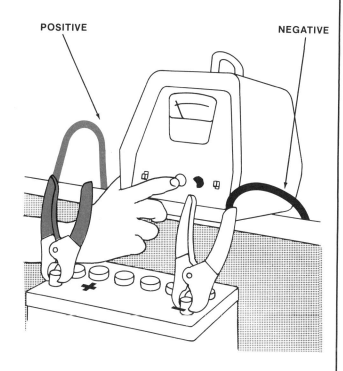

POSITIVE · NEGATIVE

11. Replace the battery vent caps and start the engine.

 CAUTION: Be sure there is plenty of fresh air when you recharge a battery. Open the doors and windows if you're inside. Hydrogen gas from the charging battery will collect in a room and a spark is all that's needed to explode it. Also, remove the vent caps when charging so gas flows freely from the battery without building up pressure inside.

HOW TO TEST A BATTERY

You should not guess how good your battery is. By following a three-step method you will know the exact condition of the battery. The three steps are:

● *Visual inspection*

● *Hydrometer test*

● *Load test*

See the following chart.

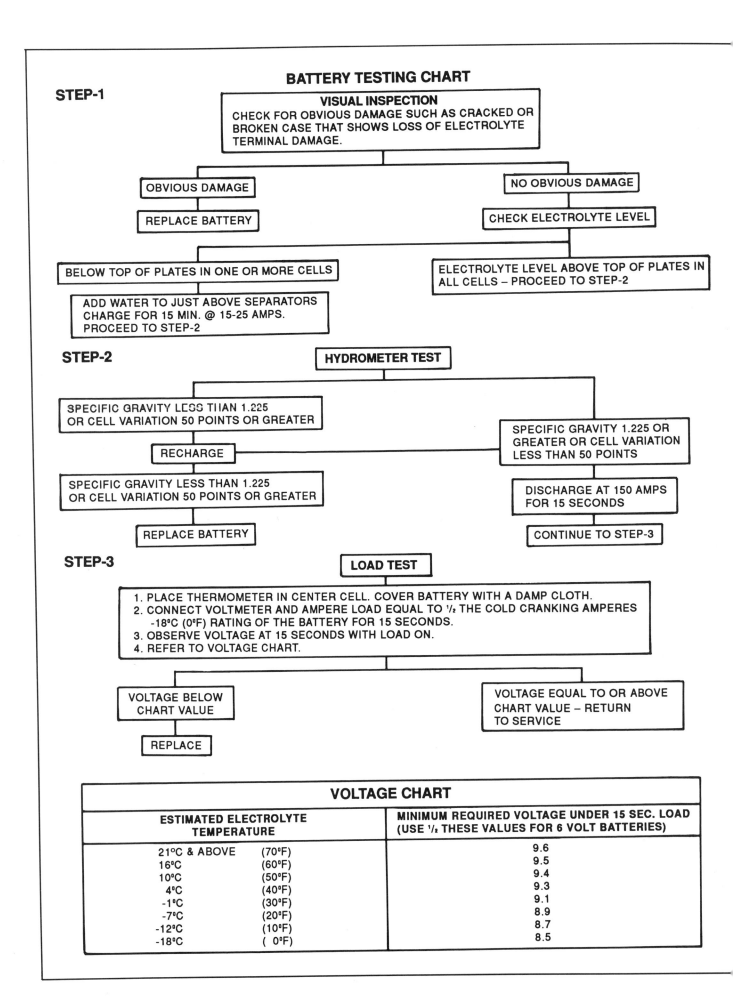

BATTERY TESTING CHART

STEP-1

VISUAL INSPECTION
CHECK FOR OBVIOUS DAMAGE SUCH AS CRACKED OR BROKEN CASE THAT SHOWS LOSS OF ELECTROLYTE TERMINAL DAMAGE.

OBVIOUS DAMAGE

REPLACE BATTERY

NO OBVIOUS DAMAGE

CHECK ELECTROLYTE LEVEL

BELOW TOP OF PLATES IN ONE OR MORE CELLS

ELECTROLYTE LEVEL ABOVE TOP OF PLATES IN ALL CELLS – PROCEED TO STEP-2

ADD WATER TO JUST ABOVE SEPARATORS CHARGE FOR 15 MIN. @ 15-25 AMPS. PROCEED TO STEP-2

STEP-2

HYDROMETER TEST

SPECIFIC GRAVITY LESS THAN 1.225 OR CELL VARIATION 50 POINTS OR GREATER

RECHARGE

SPECIFIC GRAVITY 1.225 OR GREATER OR CELL VARIATION LESS THAN 50 POINTS

SPECIFIC GRAVITY LESS THAN 1.225 OR CELL VARIATION 50 POINTS OR GREATER

DISCHARGE AT 150 AMPS FOR 15 SECONDS

REPLACE BATTERY

CONTINUE TO STEP-3

STEP-3

LOAD TEST

1. PLACE THERMOMETER IN CENTER CELL. COVER BATTERY WITH A DAMP CLOTH.
2. CONNECT VOLTMETER AND AMPERE LOAD EQUAL TO ½ THE COLD CRANKING AMPERES -18°C (0°F) RATING OF THE BATTERY FOR 15 SECONDS.
3. OBSERVE VOLTAGE AT 15 SECONDS WITH LOAD ON.
4. REFER TO VOLTAGE CHART.

VOLTAGE BELOW CHART VALUE

VOLTAGE EQUAL TO OR ABOVE CHART VALUE – RETURN TO SERVICE

REPLACE

VOLTAGE CHART	
ESTIMATED ELECTROLYTE TEMPERATURE	MINIMUM REQUIRED VOLTAGE UNDER 15 SEC. LOAD (USE ¼ THESE VALUES FOR 6 VOLT BATTERIES)
21°C & ABOVE (70°F)	9.6
16°C (60°F)	9.5
10°C (50°F)	9.4
4°C (40°F)	9.3
-1°C (30°F)	9.1
-7°C (20°F)	8.9
-12°C (10°F)	8.7
-18°C (0°F)	8.5

HOW TO START AN ENGINE WITH JUMPER CABLES AND A BOOSTER BATTERY

If a tractor has a dead battery, a second battery may be used to start the engine. The second battery, called the booster battery, may be in another vehicle. The starting procedure is:

1. **Connect the batteries**

2. **Start the engine**

3. **Disconnect the batteries**

Never connect a 12-volt battery to a 6-volt battery. It could destroy the wiring, alternator, and batteries.

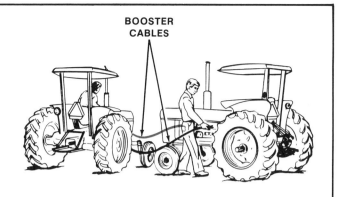

CONNECT

Stop the engine in the machine with the booster battery before connecting cables. Make sure both machines are in PARK and brakes are set. Do not let the two machines touch, as a spark can jump from one to the other and explode battery gas. Connect the cables:

1. **Connect first cable to ungrounded terminal of discharged battery.**

2. **Connect the other end of first cable to ungrounded terminal of the booster battery.**

3. **Connect one end of the second cable to the grounded terminal of the booster battery.**

4. **Connect the other end of second cable to frame of machine with discharged battery.**

Normally, red cables connect to positive (+); black cables connect to negative (−).

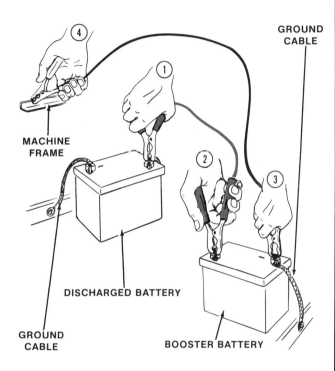

START

Start the machine that provides the booster battery and let the engine idle. Then, try to start the machine with the dead battery. If the engine doesn't start in about 30 seconds, release the starter switch, let the cables and starter cool for several minutes, then try again.

If you can't get the machine started, check the connections. Make sure the cables are connected firmly.

If you can't get the machine started after several tries and everything else seems to function, a mechanic should look at it.

DISCONNECT

After the engine starts, disconnect the cables in reverse order:

1. **Disconnect cable from frame.**

2. **Disconnect the other end from grounded terminal of booster battery.**

3. **Disconnect the cable from the ungrounded terminal of the booster battery.**

4. **Disconnect the other end.**

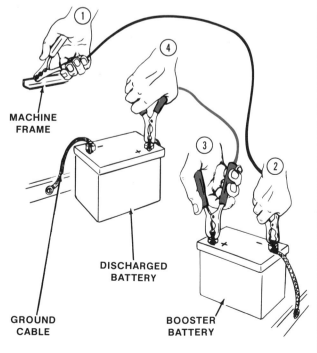

SPARK PLUG MAINTENANCE

Spark plugs make an electric spark. When they spark, they ignite the fuel and air mixture in the cylinder. Remember this from Chapter 2?

Spark plugs should be removed, cleaned, and gapped by a mechanic about every 500 hours of operation. Spark plugs have numbers on them to identify them. Replacement plugs must have the same numbers recommended in the operator's manual. These numbers indicate the heat range.

You can inspect spark plugs and avoid many problems. Look for plugs with broken or burned insulators. Burned insulators tell you the plugs need changing.

Check the spark plug wires to see if they are firmly attached to the plugs. No electricity can get to a plug if the wire is loose or broken. If you pull off the wires, remove one at a time so you don't mix them up when you put them back. Don't pull on the wires. Pull on the terminals. Pulling on the wires breaks the small wires inside.

 CAUTION: Never pull off a wire while the engine is running. You can get a bad shock.

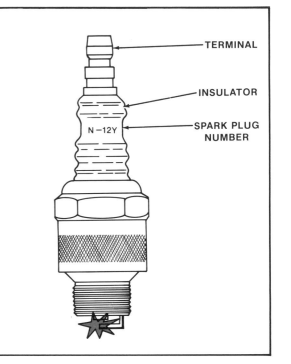

COIL, CONDENSER, AND DISTRIBUTOR MAINTENANCE

A bad coil, condenser, or distributor will cause hard starting and backfiring. The engine will overheat, waste fuel, and damage itself trying to run.

Maintenance should be performed every 500 hours of engine operation, more often if the operator's manual recommends it.

Look and listen for signs the engine isn't running properly. If you suspect trouble, avoid a breakdown by having a mechanic check the machine and make repairs.

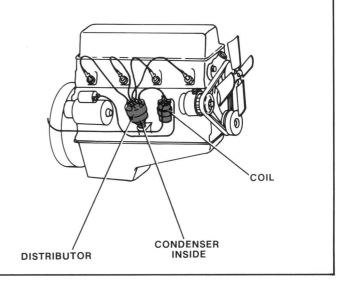

ALTERNATOR AND GENERATOR MAINTENANCE

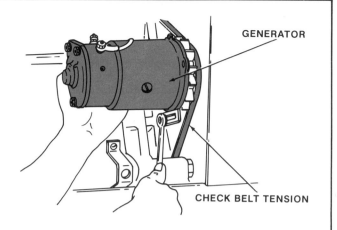

GENERATOR

CHECK BELT TENSION

Make sure wires are firmly connected to alternators and generators. Older generators must be greased according to the operator's manual.

Examine drive belts frequently. Look for frayed, cracked belts and replace them. If a belt is too loose or extra tight, adjust it.

To replace a belt:

1. Shut off engine.

2. Loosen the bolts on the alternator or generator so the belt slackens.

3. Slip the old belt off.

4. Slip a new belt on.

5. Never pry a belt off or force one on to a pulley.

Adjust the alternator or generator until the belt is fairly tight, and tighten the bolts until snug. Then, check the belt tension:

1. Wiggle the belt with your hand (engine must be shut off).

2. Pull on the belt about half way between pulleys. It should move about as far as the distance between the knuckles of your hand.

3. If the belt is too loose, pull the alternator or generator against the belt until it is tight enough.

4. Tighten bolts and recheck the tension.

5. Don't overtighten the belt. A tight belt wears and damages alternator, fan, and water pump bearings.

6. After you install a new belt, run the engine a few hours and recheck the tension. By then it should be stretched as far as it's going to.

Always have a mechanic do starter, generator, and alternator repairs. Special tools and training are necessary.

WATCH THE VOLTMETER

Look down at the gauges and signal lights often. The voltmeter, or ammeter, can tell you if the battery is being charged. After the engine has been operating for a while, the needle should indicate a low, steady rate of charge in the "Normal" range of the dial. If the gauge shows a high or low charge rate, stop the engine, find the problem, and fix it. It will probably be one of the following:

● *A loose or missing belt*

● *Loose wire connections or a damaged wire*

● *A faulty gauge; a mechanic will probably have to fix the gauge*

● *If the gauge is accurate, there may be something wrong with the alternator or generator. You need a mechanic to fix them.*

KEEP HEADLIGHTS ADJUSTED

Keep headlights adjusted so you can see rocks, bumps, and stumps at night, and so your lights don't blind others.

To adjust lights:

1. Remove the lamp mounting brackets with a wrench and aim the lights by turning the aiming screws on the mounts

2. Then, replace the mounting brackets.

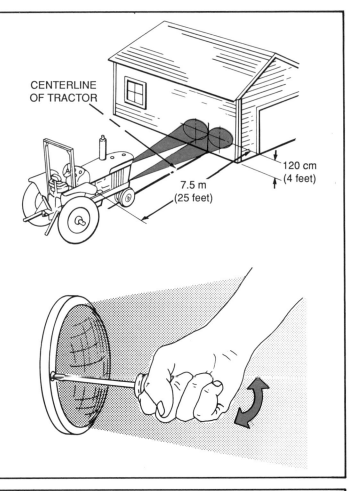

CENTERLINE OF TRACTOR

120 cm
(4 feet)

7.5 m
(25 feet)

REPLACING LAMPS AND LIGHT BULBS

Light bulbs burn out after long use or because of vibration, hard jarring, or broken switches. When a bulb burns out put a new one in promptly.

Be sure the new bulb uses the same voltage and is the same size and color as the burned out bulb. Your operator's manual will specify the kind to use. Remember, you can't take the bulb out of a sealed beam headlight. Replace the entire sealed beam with a new one.

CLEAR SHINY
GLASS

GLASS LOOKS
SMOKED

GOOD
BULB

BURNED
OUT BULB

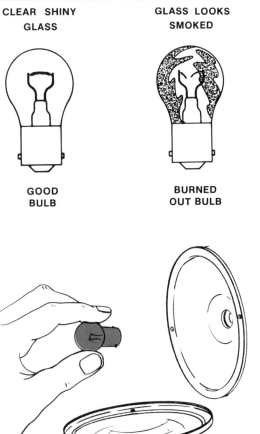

SUMMARY

The electrical system of a spark ignition engine has four circuits:

- *Starting circuit*
- *Charging circuit*
- *Ignition circuit*
- *Accessory circuits*

Each circuit has wires to carry electricity.

The electricity is generated by a generator or alternator driven by a belt from a pulley on the engine. The electricity is stored in a battery for use by the:

- *Starter motor, in the starting circuit*
- *Spark plugs in the ignition circuit (diesel engines do not have spark plugs).*
- *Instruments, lights, and accessories*

Watch for electrical system trouble signs often, just as you would look for weeds in a field. Look for signs like: low battery fluid, a loose alternator belt, worn or frayed wires, or an unusual reading on the voltmeter.

If the battery is too weak to start your machine, you can start it with jumper cables from another battery. Or, you can recharge the battery.

Fuses and circuit breakers protect electrical systems from short circuits and overloads. Fuses must be replaced if they blow. However, circuit breakers may be reset and used.

Coils, starters, condensers, and distributors should only be repaired or replaced by a mechanic.

You should be able to deflect alternator or generator drive belts about as far as it is between the knuckles of your hand.

Keep the headlights aimed and adjusted properly, and always replace burned out light bulbs promptly.

DO YOU REMEMBER

1. What are the sources of electrical power for an engine?

2. What causes short circuits?

a.

b.

3. What is a circuit breaker?

4. What causes undercharging of batteries?

a.

b.

c.

5. Name three dangers of handling battery fluid.

a.

b.

c.

6. True or False: The fluid level in most batteries only needs to be checked once a year.

7. Which cable should be taken off first when a battery is removed from a machine?

8. A discharged battery cannot be recharged.   True   False

9. How frequently should the alternator drive belt be serviced?

POWER TRAINS
CHAPTER 8

INTRODUCTION

This chapter describes:

● *Power train parts and how they work*

● *Power train maintenance*

TURNING POWER INTO WORK

Engine power can only be put to use if it is connected to the drive wheels and power take-off (PTO).

Power trains connect engine power to the drive wheels and PTO. They have the following parts:

● A **Clutch,** *to connect and disconnect power*

● A **Transmission,** *to provide several speed ranges, forward and reverse, and connect power to the PTO, on some models*

● A **Separate PTO Power Train** *is sometimes used*

● A **Differential,** *to send equal power to the final drives*

● **Final Drives,** *to carry power from the differential to the drive wheels*

Before studying these parts and their maintenance, make sure you have a clear understanding of two fundamentals:

● *Bearings*

● *How power is sent from one part to another*

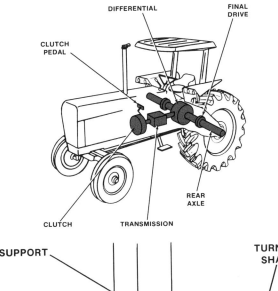

BEARINGS

Power train parts turn and carry very heavy loads. Bearings are needed to hold them in place and reduce friction. Remember this from Chapter 1?

In turn, bearings need lubrication to protect them.

HOW POWER IS TRANSFERRED

There are three basic methods of sending power from one part to another. They are:

● *Friction*

● *Gears*

● *Moving fluids*

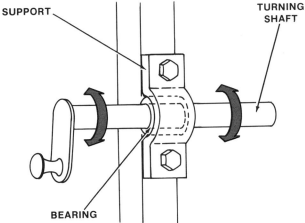

FRICTION

In the friction method, a belt from one wheel turns another. Belts are used to turn fans, water pumps, and alternators.

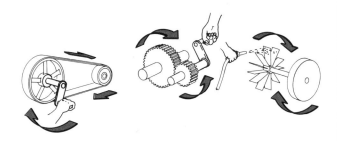

GEARS

Teeth on gears mesh together. When one gear is turned, its teeth turn the other gear. Gears are usually used in transmissions because they are small, strong, and do not slip.

FLUID

Moving fluid provides smooth power. By increasing and decreasing the pressure and rate of flow, an operator can set any speeds he wants without the jerk of changing gears. Torque converters and hydrostatic transmissions send (transmit) power with fluid.

THE CLUTCH

A clutch works like the two plates. When the plates are pressed together, the engine turns both plates, and engine power is sent through them to the transmission.

Strong springs or hydraulic pressure hold the clutch plates together. When an operator pushes down on the clutch pedal, the plates pull apart so no engine power goes to the transmission.

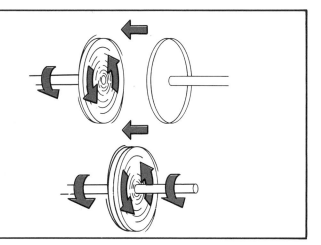

THE CLUTCH PEDAL

Letting out the clutch pedal too quickly wears the clutch plates rapidly. It also causes a sudden overload on other power train parts and can cause serious damage. On the other hand, letting the clutch pedal out too slowly lets the clutch plates slip, which wears and overheats them. "Riding" the clutch (keeping your foot on the pedal while the machine is running) lets the plates slip. This wears and overheats clutch plates fast. Riding the clutch also wears out the clutch release bearings.

A clutch pedal must move freely for a little distance before it begins to pull the clutch plates apart. This distance is called "free travel."

Free travel may vary from 13 to 50 mm (1/2 to 2 inches). You can measure the free travel. If it is more or less than specified in the operator's manual, have a mechanic adjust it. Prompt attention could reduce clutch plate wear.

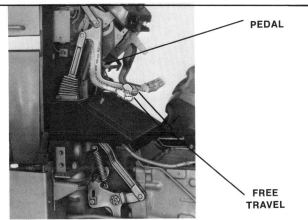

PEDAL

FREE TRAVEL

GREASING THE CLUTCH RELEASE BEARING

All clutches have a release bearing.

On some clutches the release bearing must be greased every 250 hours or so. One or two pumps of grease from a hand grease gun is enough for most release bearings. You may find the grease fitting on the side or bottom of the clutch housing. Or, you may have to remove a metal cover to reach the fitting with a grease gun. If you pump in too much grease, it can leak onto the clutch plates and make them slip, which could damage the clutch.

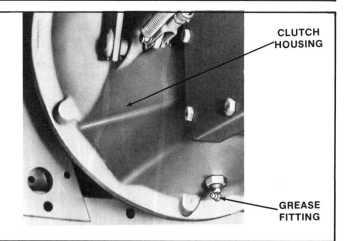

CLUTCH HOUSING

GREASE FITTING

REDUCING CLUTCH PROBLEMS

You can avoid many clutch problems by noticing change in the feel of the clutch pedal and listening as the pedal moves up and down. If you notice any of the following clutch problems, have a mechanic make repairs or adjustments immediately before problems grow.

- *Chattering or jumping, especially in low or reverse speeds, may mean there is oil, grease, or dirt on the clutch plates or parts are sticking instead of sliding smoothly.*

- *Hard shifting usually means the clutch is out of adjustment.*

- *Squeaks, especially when the pedal is depressed, can mean the clutch release bearing needs lubrication or replacement.*

- *Rattles, especially at low speeds or when the machine is standing still, can mean severe wear of clutch parts or improper adjustments.*

- *Grabbing or jerking when the pedal is released means there is oil or grease on the clutch plates.*

- *Slipping means there are worn clutch plates, weak or broken springs, or improper adjustment.*

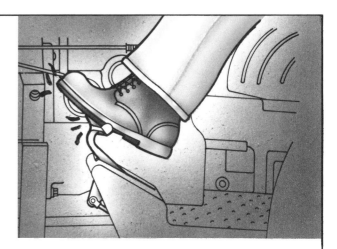

- *Vibration when you push the pedal down may mean the clutch shaft is bent.*

- *No power at the drive wheels may mean the clutch is worn out, springs are broken, adjustments are incorrect, or the clutch is not sliding on the clutch shaft.*

THE TRANSMISSION

A transmission allows an operator to increase or decrease speed and reverse the direction of travel. A typical tractor with 45 kW (60 hp) and 8 forward speeds may travel 2.5 km/hr (1.5 mph) in first gear and 26 km/hr (16 mph) in eighth gear. Tractors may have as few as 4 or as many as 20 forward gears. The lower the gear, the more weight the machine can pull, but the slower it moves.

There are four kinds of transmissions on tractors:

- *Sliding gear transmission*
- *Synchronized transmission*
- *Hydraulic assist transmission*
- *Hydrostatic transmission*

If the machine has a **sliding gear transmission,** you must stop the tractor before shifting into another gear. You will not have to stop to change speeds if there is a **synchronized or hydraulic shift transmission** in the machine. However, you still need to stop to change gear ranges. If the machine has a **hydrostatic transmission,** you never need to stop. You only need to move the speed control lever to the speed you want.

The shift linkage on some transmissions must be adjusted if the gears do not shift smoothly. Transmission repairs and adjustments require tools and training that only a mechanic has.

Transmissions can be damaged by accidents, careless operation, poor maintenance, and overloading.

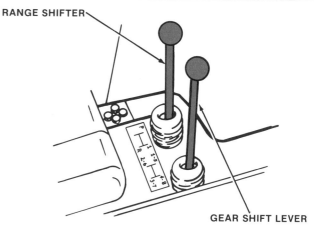

RANGE SHIFTER

GEAR SHIFT LEVER

CARELESS OPERATION:

- *Carelessly overloading the transmission, by pulling heavy loads in low gears, wears away the gear teeth.*

- *Failure to push the clutch in all the way before shifting gears may grind metal particles off the gears. The metal particles contaminate transmission oil and wear out moving transmission parts.*

POOR MAINTENANCE CAUSES:

- *Low oil level*
- *Plugged filters*
- *Overheating*

Check the transmission oil every day.

1. Turn the engine off and let the machine stand still for about 10 minutes on a level surface.

2. Check the transmission oil level with the dipstick, or

3. If the machine does not have a dipstick, remove the oil level plug and insert one finger. If the oil is up to the proper level, you should be able to feel it.

4. If oil is low, fill the transmission up to the correct level with the oil recommended in the operator's manual.

Gear oil is recommended for most standard mechanical transmissions. Automatic transmission fluid is usually recommended for hydraulic-assist transmissions. If the transmission and hydraulic system share the same oil, manufacturers usually recommend special transmission-hydraulic fluids. Using the wrong transmission oil or mixing oils of different viscosities could cause oil additives to break down, damage seals and gaskets, increase transmission wear, and plug filters.

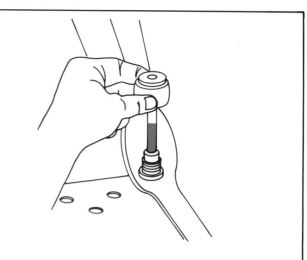

HOW TO CHANGE TRANSMISSION OIL AND FILTERS

Additives in transmission and gear oil wear out. Before they wear out, drain and refill the transmission and gear cases with the recommended oil. The operator's manual will recommend you change transmission oil every 500 to 2,000 hours. Changing oil at the recommended time for your machine will protect your machine from unnecessary wear. Write down the date and hourmeter reading when you change oil. To change oil:

1. Operate the machine long enough to warm up the oil before it is drained. Warm-up mixes dirt and sludge with the oil so it drains out with the oil.

 CAUTION: Oil may be hot and could cause burns. Let oil cool before draining it.

2. Some machines have more than one drain plug to remove to drain the oil. Be sure you remove all drain plugs or you could leave enough oil inside to make the new oil dirty.

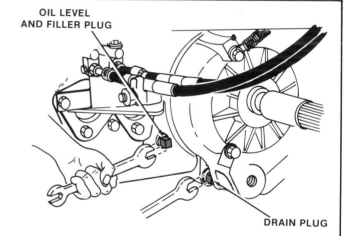

OIL LEVEL
AND FILLER PLUG

DRAIN PLUG

(continued on next page)

3. Change the transmission oil filter when the oil is changed. Use the recommended filter. Others may not provide proper protection.

4. Use filters approved or recommended by the manufacturer.

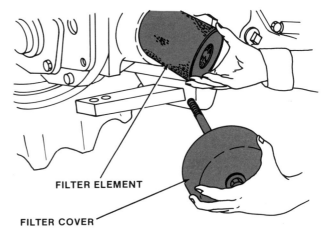

FILTER ELEMENT

FILTER COVER

5. If the transmission has a breather, remove it, clean it in diesel fuel, or a safe solvent, and reinstall it.

6. Replace and tighten the drain plugs and filter covers securely after you drain the oil.

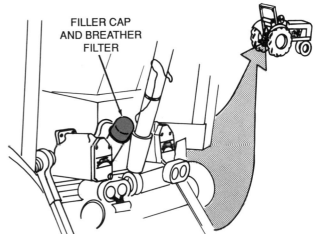

FILLER CAP
AND BREATHER
FILTER

7. Refill the system with the proper oil. But, avoid overfilling. Overfilling can make the oil foam out.

TRANSMISSION TROUBLE SIGNS

Transmission failures are expensive. They delay work and reduce yields and income. So, it pays to look, listen, feel, and smell for possible trouble signs before problems become serious failures. Be alert to sudden changes in transmission noise, vibration, and hard-shifting gears. Each time you change the oil and filter, and while you are operating the machine, look for the following signs of transmission trouble:

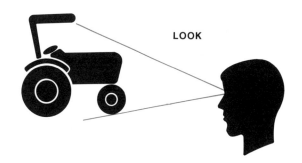

LOOK

● *Watch for oil leaks around shafts, seals, gaskets, drain plugs, filters, and other openings. Watch for cracks too. Tighten drain plugs and filters, and have other leaks repaired fast.*

● *If oil starts leaking, check the oil level more frequently.*

● *Watch gauges and signal lights for low oil pressure or high temperature.*

● *Keep the outside of the transmission clean so it can radiate heat.*

LISTEN

● *Check for water in the transmission each time you check the oil level. Water causes oil to foam and prevents proper lubrication.*

● *Listen for unusual squealing sounds from the transmission. They mean a valve is stuck. A hissing sound warns of a dry or worn bearing. A bumping sound means a bearing has worn a flat spot. Growling or scraping sounds mean worn gears.*

● *Unusual vibrations, jerky motion, and hard-shifting gears mean badly worn or damaged gears which should be replaced before they cause more damage.*

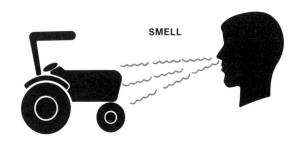

FEEL

● *Rub some oil between your fingers and feel it each time you check the oil. If the oil feels gritty, change it immediately.*

● *When you check the oil level, smell the oil too. If it smells burned, the transmission is overheating, possibly because of a plugged filter, dirty oil cooler, or low oil level. Change the oil and filter and clean the oil cooler if you smell burned oil.*

● *If you smell burned oil while operating the machine, stop immediately; find out why the transmission is overheating.*

● *If you drain the oil and find metal chips, have a mechanic check the transmission for worn or broken parts. Do not operate the machine if you find metal pieces in the oil.*

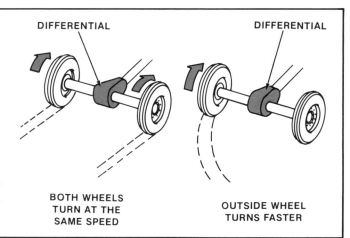

SMELL

THE DIFFERENTIAL

The differential is a box with gears inside. The gears send power from the transmission to the final drive. The differential also lets each drive wheel travel at a different speed, and still pull its share of the load. When the machine moves straight ahead, drive wheels turn at the same speed. But, when the machine turns, the wheel on the outside must turn faster than the inner wheel. The differential allows the outside wheel to turn faster as the machine turns the corner.

Usually the transmission and differential share the same oil. So, if you maintain the transmission oil correctly, you take care of the differential oil at the same time. But, check the operator's manual to be sure.

DIFFERENTIAL   DIFFERENTIAL

BOTH WHEELS TURN AT THE SAME SPEED

OUTSIDE WHEEL TURNS FASTER

FINAL DRIVES

The final drives carry power from the differential to the drive wheels. Final drives have axle shafts from the differential out to the wheels and gears to reduce speed between the differential and wheels.

Final drive gear cases are filled with oil to lubricate the gears inside. This oil should be changed when you change the transmission oil. On most tractors, the transmission, differential, and final drive all share the same oil. So, if you maintain the transmission oil correctly, you take care of the differential and final drive, too. But check the operator's manual to be sure.

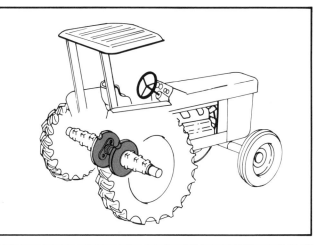

POWER TAKE-OFF

The Power Take-Off (PTO) transmits power from the tractor differential to farm implements. Tractor PTO power is carried from the tractor PTO to the implements by the rotating PTO drive shaft. The shaft has flexible joints that let the shaft keep spinning when the tractor turns a corner or goes over a bump. The shaft is made so one section can slide (telescope) in and out of the other as the implement moves up and down behind the tractor.

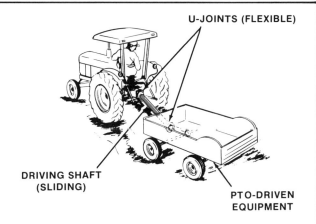

U-JOINTS (FLEXIBLE)

DRIVING SHAFT
(SLIDING)

PTO-DRIVEN
EQUIPMENT

You must grease the PTO shaft universal joints, bearings, and telescoping shaft with a grease gun every 5 or 10 operating hours.

 CAUTION: A rotating drive shaft can cause entanglement that may lead to serious injury or death. Always put all guards and shields in place.

The PTO shaft can be one of the most dangerous parts of a tractor. For your safety, a shield is placed over the rotating drive shaft. You must check to see if the safety shield turns freely before you start the tractor. If it does not spin freely, replace it.

Never remove the safety shield while working. Stand away from a **spinning** PTO shaft. And, never step over a **spinning** shaft. The shaft may catch your clothes, pull you down, break your leg, or **KILL YOU!**

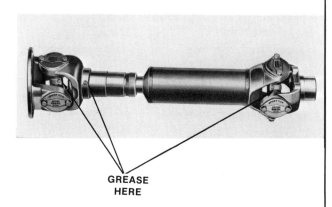

GREASE
HERE

STAY CLEAR OF ROTATING DRIVELINES

Entanglement in rotating driveline can cause serious injury or death.

Keep tractor master shield and driveline shields in place at all times except for special applications as directed in the implement operator's manual.

Wear fairly tight fitting clothing. Stop the engine and be sure power takeoff driveline is stopped before making adjustments, connections, or cleaning out PTO driven equipment.

SUMMARY

The four major parts of the power train are the:

- *Clutch*
- *Transmission*
- *Differential*
- *Final drive*

They transmit engine power to the drive wheels and PTO.

Clutch maintenance is mainly lubricating the clutch release bearing and adjusting the clutch pedal free travel.

The usual maintenance on the transmission, differential, and final drive is making sure they have adequate oil.

An operator should keep in mind that he can detect many problems and prevent damage if he is constantly looking, listening, feeling, and smelling to see if something is wrong and having it repaired before it becomes a major problem.

Since the PTO can be one of the most dangerous parts of a tractor, an operator has the job of inspecting it to be sure the main shield is in place and that the rotating safety shield turns freely and the U-joints and telescoping shaft are greased.

DO YOU REMEMBER?

1. Name four major parts of a power train.

a.

b.

c.

d.

2. List three ways power can be transmitted in a power train.

a.

b.

c.

3. Clutch release bearings require large amounts of grease at frequent intervals.   True   False

4. What are three warning signs of possible clutch troubles?

a.

b.

c.

5. What is the purpose of a transmission in a power train?

6. Gear shifting is the same on all transmissions. True   False

7. What happens if the wrong transmission oil is used?

8. List four signs of possible transmission problems.

a.

b.

c.

d.

HYDRAULICS
CHAPTER 9

INTRODUCTION

This chapter describes:

- *What hydraulic systems do*
- *How hydraulic systems work*

- *Parts*
- *Safety*
- *Maintenance*

WHAT HYDRAULIC SYSTEMS DO

Hydraulic systems send power where you want it with fluid. Remember the three ways of transmitting power in Chapter 8? When you move a control lever to raise a plow, the fluid in hydraulic lines carries your command to the plow where a hydraulic cylinder increases your strength, and lifts the plow for you.

Hydraulics do more than raise and lower plows. They also shift transmissions, pull clutch plates apart, apply brakes, help you turn the steering wheel, and more. Hydraulic systems are the "muscles" of machines.

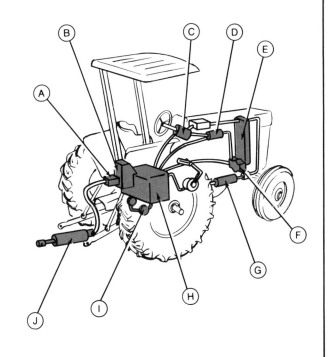

(A) COUPLER

(B) EQUIPMENT LIFT

(C) STEERING

(D) ACCUMULATOR

(E) OIL COOLER

(F) PUMP

(G) STEERING

(H) FLUID RESERVOIR

(I) BRAKE

(J) CYLINDER

HOW HYDRAULIC SYSTEMS WORK

A plastic tube in a rubber bulb full of water is a hydraulic system. If you squeeze the bulb, water squirts out the tube. The force of squeezing the bulb is sent through the liquid. If you put your finger over the end of the tube and squeeze, you can feel the force.

Similarly, when you push a hydraulic control lever, hydraulic force is sent through the hydraulic lines to power hydraulic cylinders and hydraulic motors. The motors and cylinders increase the hydraulic force and move heavy parts for you. Hydraulics is like a friend who helps you lift a heavy object.

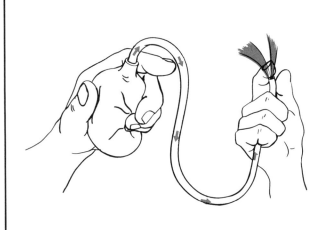

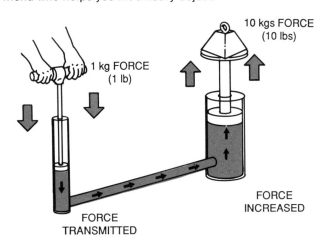

10 kgs FORCE
(10 lbs)

1 kg FORCE
(1 lb)

FORCE
TRANSMITTED

FORCE
INCREASED

HYDRAULIC SYSTEM PARTS

- *Fittings to connect lines or hoses to parts*
- *Pumps to make pressure (a force) in the system*
- *Valves to direct and adjust the pressure*
- *Cylinders to turn the hydraulic pressure into a push or pull*
- *Motors to turn the hydraulic pressure into rotary power*
- *Oil lines or hoses to carry the hydraulic oil between the pump, valves, cylinders, etc.*
- *Coolers to cool the hydraulic oil*
- *Oil to carry the pressure through the system*

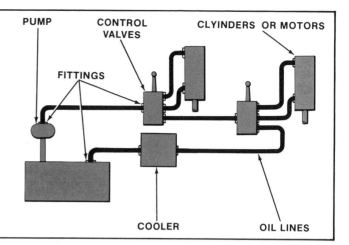

HYDRAULIC FITTINGS

Hydraulic fittings connect lines or hoses to other parts. They must be tight enough not to leak, but not overtightened. Small leaks can lower oil pressure, cause a fire, let in dirt, and cause the system to heat up and damage itself. A small leak can grow into a big one fast. If you see a loose fitting, tighten it. You may need to replace it or a seal in it if it will not stop leaking. If you replace it yourself, follow the safe procedure:

1. Turn off the engine, put the machine in PARK, set the brakes, and block the wheels.

2. Relieve system pressure by moving the control levers back and forth.

3. Loosen fittings slowly and carefully.

4. Do not let oil squirt on you.

5. Replace the damaged fitting with the recommended fitting only.

Hydraulic fittings come in a great variety because they all connect to very specific areas, such as ports, lines, hoses, valves, cylinders, motors, and etc. You must use extreme care in replacing fittings to make sure they have the right thread, because they could be metric or inch dimensioned, or have

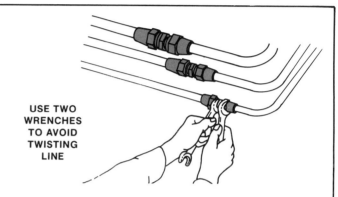

USE TWO WRENCHES TO AVOID TWISTING LINE

straight or tapered threads. If you use the wrong fitting, you may have leakage or do damage to other parts. Fittings seal in different ways, such as thread interference (pipe thread, NPT) or metal-to-metal seal (37 degree flare or 24 degree cone), or O-ring seal.

IMPORTANT: Never overtighten a fitting. Just tighten it enough to seal. Also prevent pipe thread sealant from entering the hydraulic system.

HYDRAULIC PUMPS

The hydraulic pump pumps a volume of pressurized oil into the hydraulic system that the cylinders and hydraulic motors use. Many small, delicate parts are inside hydraulic pumps. They can be damaged by:

- *Dirt, rust, metal particles, and even bubbles*

- *Sludge formed by heating dirt and oil in the pump*
- *Bad hydraulic oil that is too thick, too thin, or has the wrong additives*
- *Lack of oil*
- *Overloading the pump*

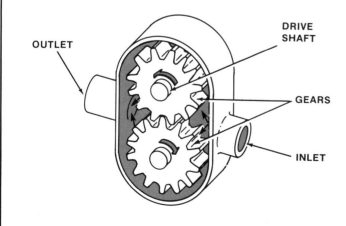

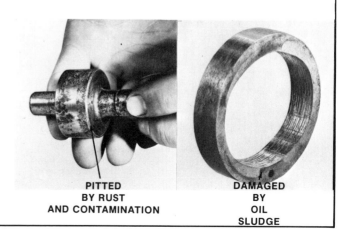

PITTED BY RUST AND CONTAMINATION

DAMAGED BY OIL SLUDGE

HYDRAULIC VALVES

Valves are little gates that can be opened and closed to control the flow of hydraulic oil. There are three main kinds of valves:

• **Pressure control valves** *can open or close to keep pressure from rising too high or falling too low in the system.*

• **Directional control valves** *can be opened and closed to send pressure to hydraulic cylinders or motors.*

• **Volume control valves** *can be opened or closed to increase or decrease the volume of fluid flowing through the system.*

Valves are delicate. If dirt, rust, or metal particles get into the hydraulic oil, they can easily scratch close-fitting valve parts or hold the valves open. Water in the oil can rust valves. Even lint, from a dipstick cloth, may catch in a valve and hold it open. Oil is also important. If you use oil that is too thin, it may leak through valves; thick oil may plug valves and cause slow action in the hydraulic system.

To keep dirt and water out of the system:

• *Tighten leaking fittings*

• *Replace leaking lines*

• *Use the recommended oil*

• *Operate valves carefully so you do not damage them*

• *Clean the fittings before you hook up an implement to the tractor hydraulic system.*

Most valve repairs require special tools and training. Always rely on mechanics for valve repair.

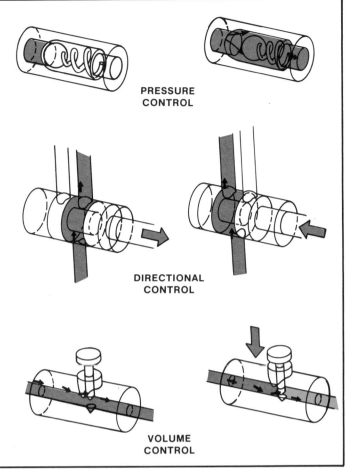

PRESSURE CONTROL

DIRECTIONAL CONTROL

VOLUME CONTROL

HYDRAULIC CYLINDERS

Hydraulic cylinders change hydraulic oil under pressure into push or pull motion. Double-acting cylinders push the rod out and pull it back in. Single-acting cylinders just push the rod out. The weight of a load must push the rod back in.

Dirt in hydraulic oil is the main problem. Dirty oil can cut seals, scratch cylinder rods, and cause leaks.

You can protect the hydraulic system from dirt if you:

• *Store cylinders fully retracted so dirt cannot stick to the exposed rod.*

• *Keep vents wiped clean.*

• *Keep fittings tight.*

• *Keep cylinders aligned. If they are misaligned, they can twist, unseat, or crush seals, and bend or break the cylinder rod.*

• *Clean the fittings before you hook an implement up to the tractor hydraulic system.*

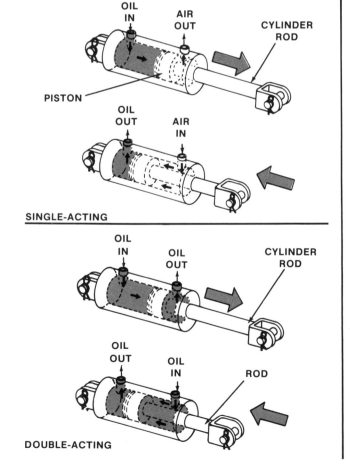

SINGLE-ACTING

DOUBLE-ACTING

(continued on next page)

If the cylinder action is "spongy," air is probably trapped in the system. You can bleed it out as follows:

1. Put the tractor in PARK with the engine running, the brakes set, and wheels blocked.

2. Connect the hydraulic hoses from a cylinder to the tractor.

3. Stand the cylinder on end with the rod down on the ground.

4. Support the cylinder. Extend and retract the cylinder rod a few times in order to bleed trapped air.

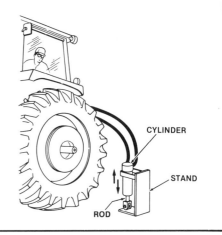

HYDRAULIC MOTORS

Hydraulic motors change oil under pressure into rotary motion to turn gear and pulley drive shafts.

Again, dirt and water are the main problems. If they get into the oil, they will ruin a motor.

Hydraulic motors can also be damaged if the mounting bolts are loose or the motor is not in line with the shaft it turns. Check the motor alignment often, and keep the bolts tight.

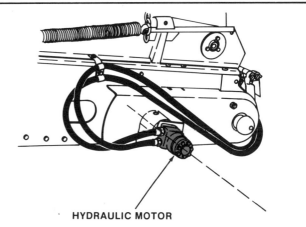

HYDRAULIC LINES AND HOSES

Hydraulic hoses connect all parts of a hydraulic system. Hoses are very strong to withstand oil pressure and chemical action. But, they can wear out.

You often see small cracks on the surface of hydraulic hoses. Small cracks do not always mean the hose must be replaced. However, they lead to larger cracks and pinhole leaks that can stop your machine.

You can avoid big problems by carefully inspecting hoses. See "How to Find Leaks."

If you find a leaking or damaged hose, replace it immediately. But follow safe procedures:

1. Turn off the engine, put machine in PARK, set the brakes, and block the wheels.

2. Relieve pressure by moving the control levers back and forth.

3. Loosen fittings slowly and carefully.

4. Do not let oil squirt on you.

5. Replace the damaged line or hose with the recommended kind of line or hose only.

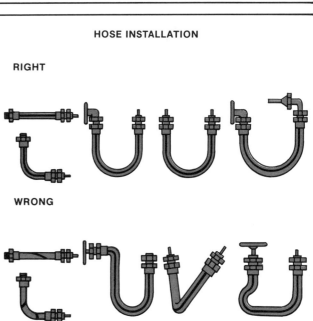

HOSE INSTALLATION

RIGHT

WRONG

OIL COOLERS

On machines with many pieces of hydraulic equipment, the oil may overheat unless it is cooled. Overheating thins the oil, shrinks seals and gaskets until they leak, and coats parts with a hard oil varnish.

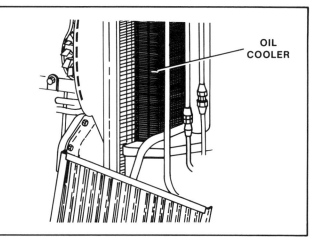

OIL COOLER

Oil coolers must be kept clean.

1. **Wipe the dirt and chaff off every day.**

2. **Fix leaking connections.**

3. **Use recommended oil at correct level.**

4. **Keep the belt on the oil pump tensioned correctly.**

5. **Have a mechanic drain, flush, refill, and bleed the oil system if the hydraulic oil is dirty.**

HYDRAULIC OIL

Hydraulic oil must flow when it is hot or cold. It must lubricate moving parts and protect parts from rust and corrosion. It must not wear out during hours of tough work. It must also resist foaming. Machine manufacturers recommend the kind of hydraulic oil that will do these jobs best in their machines. Always use the recommended oil.

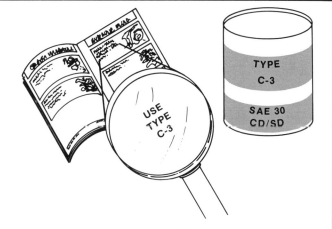

USE TYPE C-3

TYPE C-3

SAE 30 CD/SD

Remember most hydraulic problems happen because water and dirt are allowed to get into the oil. Water and dirt usually come from:

- *Moist air entering the breather cap and condensing inside*

- *Dirt and chaff falling in when the filler cap or dipstick is removed*

- *Dirt squeezing by leaking seals and gaskets*

- *Metal particles worn from moving machine parts*

- *Dirty hydraulic hookup fittings*

You can keep the water and dirt out if you:

- *Wipe dirt off the dipstick and dipstick areas when you check the oil*

- *Wipe dirt off oil cans and spouts before pouring in oil*

- *Have leaks repaired promptly*

- *Clean hookup fittings*

If you find metal particles in the oil, it's a sign of major wear. Have a mechanic repair the problem fast.

HYDRAULIC SYSTEM SAFETY

Even though safety precautions were listed before, they are worth repeating. Good maintenance reduces the chance of hydraulic failure. But, maintenance is not worth using dangerous methods that could hurt you or others. Do not make mistakes. Follow safe practices.

⚠️ **CAUTION: Hydraulic systems contain oil under high pressure. Some systems also have accumulators that store energy even after the engine is shut off.**

Before working on the hydraulic system you must shut off the engine. Then move each hydraulic control lever or pedal several times to relieve the system pressure.

TURN OFF THE MACHINE, PUT IT IN PARK, AND BLOCK THE WHEELS BEFORE WORKING ON HYDRAULIC SYSTEM.

(continued on next page)

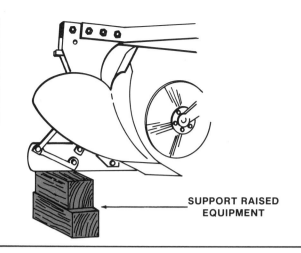

SUPPORT RAISED EQUIPMENT

THE FOUR ENEMIES OF HYDRAULIC SYSTEMS

Guard the system against the four basic enemies that attack hydraulic systems:

- *Low oil level*
- *Dirt*
- *Leaks*
- *Incorrect oil*

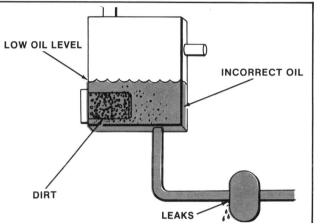

LOW OIL LEVEL

INCORRECT OIL

DIRT

LEAKS

HOW TO CHECK HYDRAULIC OIL

Park the machine on a level spot so you can get an accurate reading. The operator's manual will tell whether the engine should be running or stopped when oil level is checked.

Wipe dirt away from the dipstick area. Pull out the dipstick, or look in the sight glass if there is one. Look at the oil level on the stick or glass, and look for bubbles, foam, water, and dirt. Oil will foam if there isn't enough in the system. Water makes oil milky. If you must add oil frequently, look for leaks around fittings and seals and cracks in the reservoir and cylinders. If dirt and water are in the oil a mechanic must drain, flush, fill, and bleed the system.

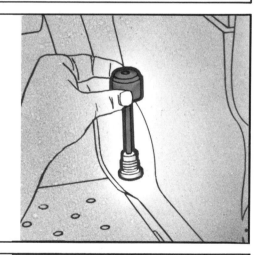

HOW TO ADD OIL

If you find the hydraulic oil is below the full line on the dipstick, or the line on the sight glass, add oil as follows:

1. Shut off the engine, put the machine in PARK, set the brakes, and block the wheels.

2. Clean the area around the filler cap.

3. Clean the tops of oil cans plus spouts or funnels.

4. Fill the reservoir with the recommended oil and re-place the filler cap.

5. Start the engine and warm it up to normal temperature.

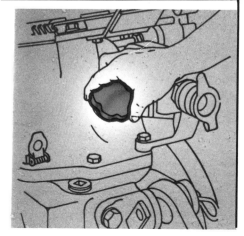

(continued on next page)

6. Then, operate all the hydraulic controls four to five times.

7. Stop the engine and recheck the oil level.

8. Add more oil if it still isn't up to the full line on the dipstick or sight glass.

NOTE: Always check the oil level after making hydraulic system repairs.

HOW TO CHANGE FILTERS

A clean filter catches and holds dirt, rust, and metal particles. But, the filter can only hold a certain amount of these contaminants. When it is full, it must be changed or dirty oil will be pumped through the system.

1. Change the filter with the engine shut off, machine in PARK, brakes set, and wheels blocked.

2. Remove the filter cap or screw-on filter element.

3. Slide out the old filter.

4. Wash the inlet screen with diesel fuel.

5. Slide in a new filter.

6. Replace the filter cap and snug it down so there is no leak.

Always change hydraulic filters at the time recommended in the operator's manual. Use recommended filters only. Different filters may not do the job.

Don't let dirt fall in when you change the filter. Be careful handling the filter, too. You can easily dent a filter if you shove it in hard. And, when you change the hydraulic oil and filter, wash the oil pump inlet screen with diesel fuel.

NOTE: If there are dirt or metal particles in the oil, have a mechanic check the system immediately.

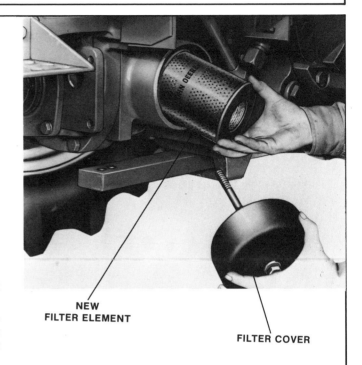

NEW
FILTER ELEMENT

FILTER COVER

HOW TO FIND LEAKS

You must leave the engine running to find pinhole leaks in hoses and fittings. Stay clear of moving parts.

1. With the engine running, place the machine in PARK, set the brakes, and block the wheels.

2. Look for leaks with a piece of cardboard, not your hand.

3. Turn off the engine and relieve pressure before you make repairs.

You can also check for leaks if you raise an implement, put the valve levers in neutral, and stop the engine. Watch to see if the implement stays up or lowers to the ground. If it lowers, there is probably a leak in the cylinder, hoses, fittings, or valves. If you can't find and fix the leak, have a mechanic do it.

Cylinders that respond slowly or move without you touching the control lever may have leaks. Outside leaks are easy to see. The leaking oil collects dirt and chaff. Leaks inside cylinders are not so obvious; but they are just as important. Leaks waste oil and may suddenly rupture causing loss of control, severe burns, and eye injuries. If you suspect a cylinder of leaking, replace it or have it repaired.

If the hydraulic reservoir oil is lower than usual, check all the lines, cylinders, and valves for leaks.

(continued on next page)

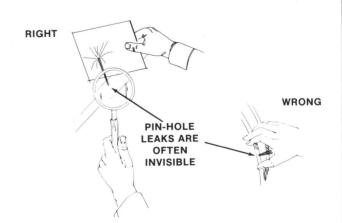

RIGHT

WRONG

PIN-HOLE
LEAKS ARE
OFTEN
INVISIBLE

AVOID HIGH-PRESSURE FLUIDS

 CAUTION: Escaping fluid under pressure can penetrate the skin causing serious injury. Avoid the hazard by relieving pressure before disconnecting hydraulic or other lines. Tighten all connections before applying pressure. Search for leaks with a piece of cardboard. Protect hands and body from high pressure fluids.

If an accident occurs, see a doctor immediately. Any fluid injected into the skin must be surgically removed within a few hours or gangrene may result. Doctors unfamiliar with this type injury should contact other knowledgeable medical sources.

HYDRAULIC POWER STEERING

Hydraulic power makes it much easier for you to steer. When you turn the steering wheel you open valves and oil flows to a hydraulic cylinder that steers the wheels for you.

The main maintenance jobs are the same as those for other hydraulic systems:

- *Make sure there is enough oil*
- *Use the recommended oil*
- *Keep the oil clean*
- *Fix leaks before they grow*

Some machines have a separate power steering pump driven by a belt from the engine. If your machine has one, be sure you check the belt tension frequently. You should be able to wiggle it about as far as between the knuckles of your hand. If it moves too little or too much:

1. Shut off the engine, put the machine in PARK, set the brakes, and block the wheels.

2. Loosen the hold down bolts on the oil pump; belt tightener if your machine has one.

3. Move the pump or tightener until you get the right belt tension.

4. Replace the belt with the recommended type if it is frayed or cut.

5. Retighten the bolts

If steering problems continue, have a mechanic examine the steering system.

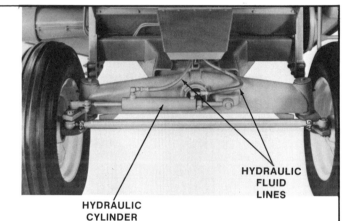

HYDRAULIC CYLINDER

HYDRAULIC FLUID LINES

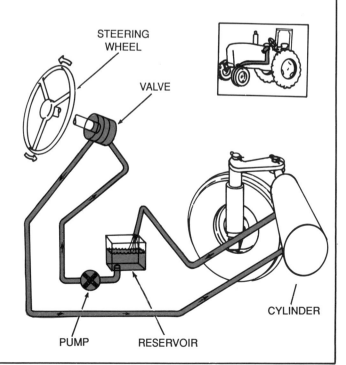

STEERING WHEEL

VALVE

CYLINDER

PUMP   RESERVOIR

DIPSTICK

PULLEY

HYDRAULIC POWER BRAKES

- *Check the brakes **before** you get out in the field by test-stopping a couple of times.*
- *Look for leaks and fix them.*
- *Make sure the oil level is correct.*

Brakes should feel firm when you push on the pedals. If the brakes feel soft or spongy, there is probably air in the hydraulic lines. A mechanic can remove the trapped air by bleeding the system. The mechanic will open bleed screws and push on the brake pedals to force the trapped air out past the loosened screws. If bleeding does not correct the spongy feel, mechanical repairs are probably needed.

If your brakes grab, squeak, or slip, they may be badly worn or out of adjustment. Have them repaired and adjusted by a mechanic.

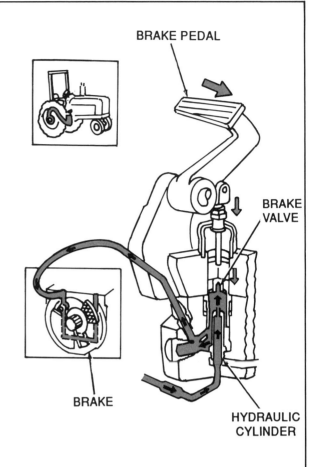

BRAKE PEDAL

BRAKE VALVE

BRAKE

HYDRAULIC CYLINDER

HYDRAULIC CONTROLLED 3-POINT HITCHES

There are four sizes of hitches: type I, type II, type III, and type IV. Type I is for small tractors and implements. Type IV is for the largest machines.

Tractor 3-point hitches may be controlled manually (depth or position control) or automatically (load or draft control). Hitch controls raise or lower the implement to maintain a constant load on the tractor.

The load system may respond too quickly to changes in load making the hitch "chatter" up and down. If this happens, adjust the response control to slow the hitch action. Set the system for faster response if the hitch responds too slowly.

Grease and adjust 3-point hitches when recommended in the operator's manual. Rely on a mechanic to make major adjustments, or to check the hydraulic system.

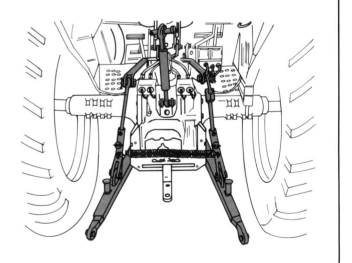

REMOTE IMPLEMENT HYDRAULIC SYSTEM

Hydraulic power is often used to raise, lower, and control pulled implements or equipment mounted on a tractor. In the normal arrangement there are:

- *Control valves on the tractor*
- *Hoses from the hydraulic couplings to the implement*
- *Hydraulic cylinders or motors on the remote implement*

Dirt is the main problem. It can be kept out if you:

- *Put the dust plugs in the tractor hose couplings when you aren't using them.*
- *Cover the hose ends when they aren't being used.*
- *Wipe the couplings and hoses off with a clean cloth before you connect them.*
- *Store cylinders with the rod retracted so dirt can't collect on it, or cover it with grease that you can wipe off before the next use.*

SUMMARY

Hydraulic cylinders and motors are the muscles for machines. But, they can be crippled by leaks, dirt, water, lack of oil, improper oil, and overloading.

All the parts of most hydraulic systems share the same oil. This means they also share the same dirt if you let it get in.

Leaks let in dirt and are dangerous. You should look for leaks constantly and fix them before they stop your machine or injure you.

The four main enemies of hydraulic systems are:

- *Low oil level*
- *Dirt*
- *Leaks*
- *Incorrect oil*

When you stop a machine to fix a leak or to inspect, follow these safe practices:

- *With the engine running, put the machine in PARK, set the brakes, and block the wheels.*
- *Check for leaks with a piece of cardboard so the oil squirts on it, not you.*
- *Stop the engine before you make repairs.*
- *Relieve pressure by working control levers back and forth. Then make repairs.*
- *Always loosen fittings slowly so oil is controlled.*

DO YOU REMEMBER?

1. Small gates in a hydraulic system, which open and close, are called _____.

2. Hydraulic pressure can be push and pull pressure if a _____ is used.

3. Dirt is the number one enemy of hydraulic systems. True   False

4. Hydraulic oil that is foaming indicates that the operator should _____.

5. Oil filters can be cleaned.   True   False

6. How do you safely find pinhole leaks in a hydraulic system?

7. Why is hydraulic power used to control steering, brakes, and other parts of a machine?

INDIVIDUAL PARTS
CHAPTER 10

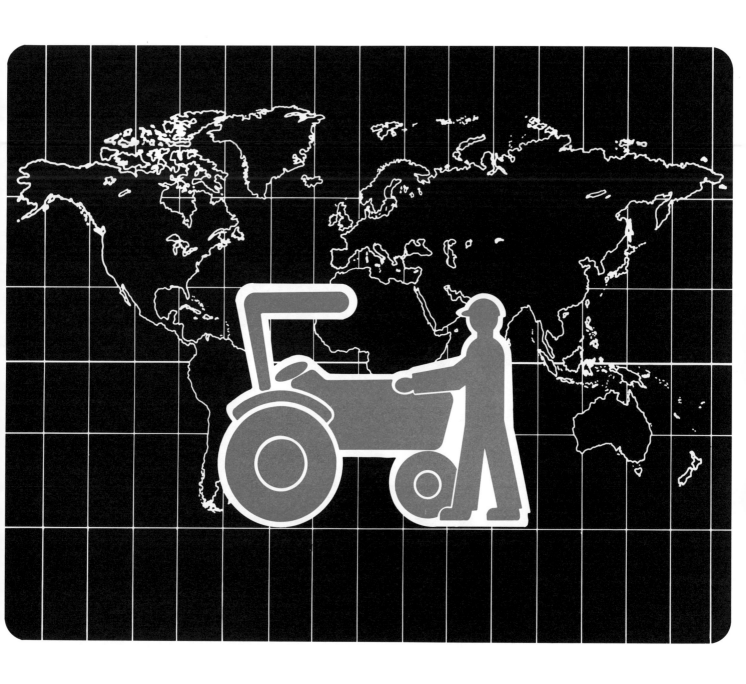

INTRODUCTION

This chapter describes the special care required for:

- *Tires, wheels, steering, and brakes*

- *Belts, chains, and gears*

- *Lubricating and cleaning tractors and other machines*

TIRES

Machines like tractors and combines need the right size tire for their weight and the soil. Small tires on a heavy machine can spin easily in soft soil. They may not have the strength and will blow out. Oversized tires waste fuel and put extra stress on machines.

Machines also need the right tire tread. Tires with the familiar herringbone bar tread are standard on most tractor drive wheels. But, deeper-treaded tires are made to grip extremely soft soil like you would find in rice or sugarcane fields. Some tires are made with criss-cross cuts instead of tread bars so they won't leave deep tracks in lawns. Industrial lug tires are designed to work either in the field or on pavement.

Tires with one, two, or three ribs are generally used on tractor front wheels which turn to steer the machine. Single-ribbed tires give the best control, but they cut deeply into soil. So, three-ribbed tires are most common.

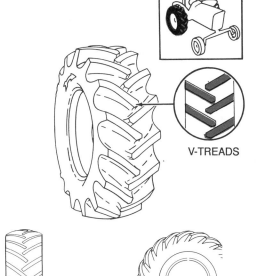

V-TREADS

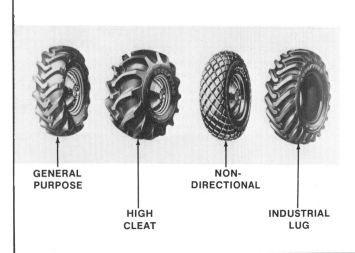

GENERAL PURPOSE

HIGH CLEAT

NON-DIRECTIONAL

INDUSTRIAL LUG

TIRE INFLATION

When tires are inflated to the correct pressure, they:

- *Grab the ground for better traction*

- *Don't slip on the rim and cut the valve stem*

- *Float over the ground instead of sinking in*

- *Hold up heavy loads without buckling*

- *Last about 25 percent longer than improperly inflated tires*

Overinflated tires rapidly wear away the center of their tread, and are so hard they are easily damaged by rocks and stumps.

Underinflated tires buckle, overheat, slip on the rim, and wear out quickly.

You can measure the air pressure in tires with a small hand-held pressure gauge or a big gauge built into an air hose chuck. Your operator's manual will recommend the correct tire pressure. You measure the tire pressure to see if it is correct.

- *Use a tire gauge you know is accurate.*

- *Don't use a gauge made for measuring high pressure to measure low pressure tires.*

- *Measure pressure when the tires are cool.*

(continued on next page)

- *If the tire contains liquid ballast, measure the pressure when the wheel is turned so that the valve stem is at the lowest point on the wheel. If the valve stem cannot be positioned at the bottom when pressure is measured, then add to the reading: 3.5 kPa for each 30 cm (1 psi for each foot) the valve stem is above the lower rim.*

- *Use a liquid pressure gauge if the tire has liquid ballast.*

- *Remember to wash off a liquid pressure gauge when you are through.*

 CAUTION: If you add air to a tire, put a little air in at a time. Then recheck the pressure. Also:

- **Do not inflate tires beyond the maximum pressure rating (usually printed on the side). (Tires can BLOW UP with enough force to injure you FATALLY if over inflated.)**

- **When inflating tires, use a clip-on safety chuck and extension hose long enough to allow you to stand to one side and NOT in front of or over the tire assembly. Use a safety cage if available.**

- If you have a tire with low air pressure:

- *Inflate it to the pressure that is recommended in the operator's manual.*

- *If the tire has lost air, it probably has a leak. Have it repaired before using the machine.*

OVER INFLATION   UNDER INFLATION   PROPER INFLATION

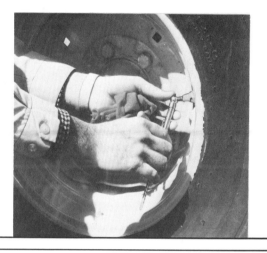

DRIVE WHEEL BALLAST

There are two kinds of drive wheel ballast: metal and liquid. Metal ballast weights are convenient because you can add or remove a few to adjust the ballast. However, they are expensive. Liquid ballast is cheaper, but liquid ballast is more permanent. A trained tire serviceman can measure the tire slip and pump in the right amount of liquid ballast to control the slip. He will also pump in calcium chloride so the water won't freeze.

Drive wheels must have enough weight to grip the soil. But tractor weight alone may not be enough to pull heavy loads. More ballast or weight may be required. Too much weight wastes power and fuel. If wheels don't have enough weight, tires slip too much and wear out too fast. You can look at tire tracks when the tractor is pulling a load and see if more ballast is needed. **Too much weight** leaves sharp, clear tracks with no signs of wheel slip. If there is **too little weight**, tracks are scratched out. When tires have the **right weight**, soil between tread marks is shifted, but the tread marks are still clear.

 CAUTION: Don't add too much weight. Each tire is marked for the biggest load it can carry. Putting on too much ballast may overload the carrying capacity of the tire and cause it to blow-out.

AIR COMPRESSES LIKE A CUSHION   WATER CAN'T BE COMPRESSED

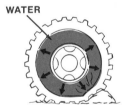

AIR

WATER

WATER

CORRECT - 75% FULL   INCORRECT - 100% FULL

FRONT BALLAST

About a third of a tractor's weight should be on the front wheels for good steering. (Four-wheel-drive tractors and self-propelled harvesting machines are exceptions.)

Iron weights are attached to the front of tractors if there isn't enough weight on the front wheels. The weights are made in standard sizes for convenience.

DUAL WHEELS

Dual wheels usually give you better traction and don't sink into soft soil as much as single wheels. However, they may overload the axles, bearings, and power train on some tractors. So, before adding duals, be sure your tractor is designed to absorb the additional stress. Check your operator's manual.

Inside and outside dual tires must have the same overall diameter to avoid tire and machine damage. If the outside diameter of one pair of tires is larger than the other, mount the big ones on the inside.

TIRE FAILURES

Avoid tire problems. Keep tires properly inflated, and don't run over stumps and sharp rocks. If small cuts and punctures are repaired promptly, tires will last longer. However, if a cut is too big to repair, don't delay. Replace the tire.

 CAUTION: Tractor tires are heavy and hard to handle. You need special equipment to repair them. Don't try to remove or repair tires unless you have the right training and equipment. You could be hurt and the tires could be damaged beyond repair. Tire repairs should be done by trained tire servicemen.

Wash tires often so accumulated grease, oil, and chemicals don't weaken the rubber. Sunlight damages tires, too. So it's best to store machines inside or cover the tires.

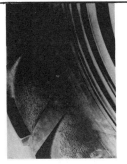

WEATHER CHECKS ON THE TIRE ARE NOT USUALLY HARMFUL

WEAR DUE TO SPINNING TIRES

STUBBLE WEAR (FARM TRACTORS)

TREAD WIPING FROM OPERATION ON HARD SURFACED ROADS

(continued on next page)

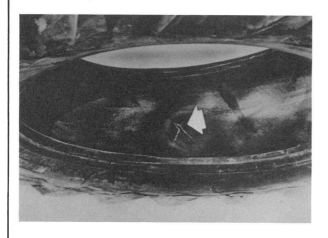

THIS SMALL FABRIC BREAK CAN BE REPAIRED
AND THE TIRE KEPT FOR ITS FULL SERVICE LIFE.
IT IS TYPICAL OF THE AVERAGE SMALL BREAK
IN A REAR TRACTOR TIRE.

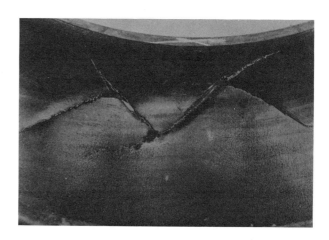

FREEZING RUINED THIS TIRE. FILLED WITH
WATER, IT WAS BROKEN WHEN THE WATER
FROZE AND EXPANDED. ANTIFREEZE
SOLUTION WOULD HAVE
PREVENTED THIS.

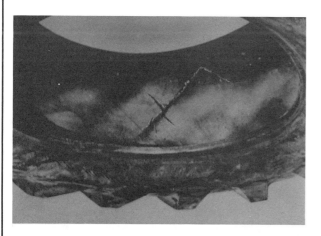

THIS BREAK WAS CAUSED BY TIRE HITTING
SOME OBJECT. THE FABRIC WAS UNABLE TO
WITHSTAND THE SHOCK.

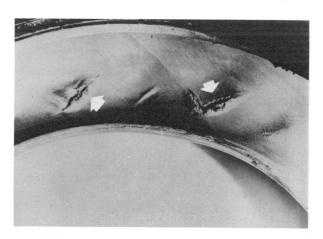

FURROW TIRE ON FARM TRACTOR WHICH HAS
LOW-PRESSURE BUCKLE OR FURROW BREAK.

FURROW TIRE WHICH HAS SIDEWALL
CRACKING DUE TO UNDERINFLATION AND
HEAVY DRAWBAR LOAD.

MANUAL STEERING ADJUSTMENT

Check the oil level in the manual steering gear case every 250 hours; sooner if recommended in the operator's manual. If the oil is low:

1. Fill the gear case with the recommended oil.

2. If the case leaks, tighten the gear case bolts.

3. If it still leaks, have the seals replaced by a mechanic.

After long use, the manual steering gear may develop looseness. You can tell because you can turn the steering wheel back and forth 50 mm (2 inches) without turning the front wheels.

If the wheel is loose, have it adjusted; or adjust it yourself as follows:

1. Turn the steering wheel all the way from left to right, counting the turns of the steering wheel.

2. Then turn the wheel back half the number of turns. The wheels should be pointing straight ahead now.

3. Loosen the lock nut on the steering gear adjusting screw with a wrench.

4. Then, tighten the adjusting screw with a screwdriver.

5. Hold the adjusting screw with a screwdriver so it will not turn, and retighten the lock nut with a wrench.

6. Check your adjustment. See if you can turn the steering wheel 50 mm (2 inches) without turning the front wheels. If you can, repeat the procedure.

NOTE: If there is still looseness it may be due to worn steering linkage.

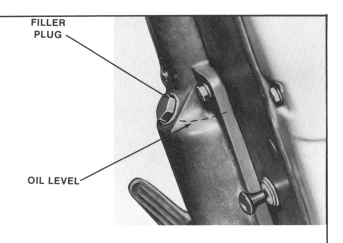

FILLER PLUG

OIL LEVEL

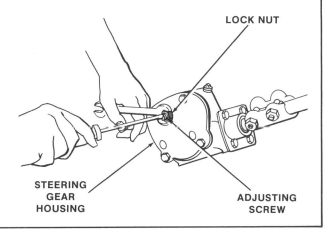

LOCK NUT

STEERING GEAR HOUSING

ADJUSTING SCREW

FRONT WHEEL TOE-IN

The front wheels of tractors, and some other machines, must turn in (toe-in) slightly for good steering control and so the tires last longer.

Check the toe-in once a year, if steering becomes hard, or if tires show wear on the inside. Always check toe-in after you run into something.

1. Drive your machine out on level ground, and shut off the engine. Let it roll to a stop without using the brakes. The wheels should be pointing straight ahead. Put the machine in PARK and set the brakes.

2. Now measure the distance between the front tires. Measure it twice; once across the front of the tires and again across the rear. Measure at hub level.

3. The distance should be 3 to 10 mm (1/8 to 3/8 inch) less across the front of the tires than across the back.

4. If the toe-in is wrong on a machine with one tie rod, loosen the tie rod clamps and turn the tie rod until the toe-in is correct. Then retighten the clamps.

5. On a machine with two tie rods, loosen the clamps on each tie rod. Turn each tie rod an equal amount until the toe-in is correct. Then retighten the clamps.

NOTE: Both wheels must have equal toe-in or steering will be hard, and the tires will wear quickly.

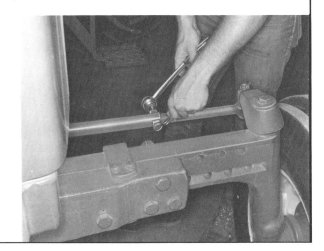

WHEEL BEARINGS

You should usually clean bearings and pack them with grease every 1,000 hours or once a year. But, some wheel bearings need grease as often as every 10 hours. Read the operator's manual to see what is recommended for your machine. Clean and pack as follows:

1. Shut off the engine, put the machine in PARK, and set the brakes. Then jack up the machine so the front wheels are off the ground. Then, slide in blocks or steel support stands to hold up the machine.

 CAUTION: Never use jacks alone to support the machine. Always have blocks or steel stands under the machine. Do not use concrete blocks! They crack and fall apart easily.

2. Remove the hub cap.

3. Pull the cotter pin out of the slotted nut.

4. Remove the slotted nut, washer, wheel hub, and bearings.

5. Wash all the dirt and grease off these parts with diesel fuel.

6. Dry the bearings carefully. Use dry, compressed air if you can. Never spin the dry bearings; that scratches them. They wear fast if you scratch them.

7. Inspect the bearings for scratches, pits, and flat places. Replace damaged bearings. Replace damaged seals while you are at it. Make sure you replace parts with identical ones recommended in the operator's manual.

8. Pack the bearings thoroughly with the kind of grease recommended in your operator's manual.

9. Reassemble all the parts exactly the way they were.

10. Tighten the slotted nut until you can't wiggle the front wheels. Grab the wheel with both hands and try to wiggle it to check.

11. Use a torque wrench to tighten the slotted nut to about 48 N•m (35 lb-ft) or other torque specified in the operator's manual. Then, turn the nut back to the nearest slot and insert the cotter pin.

12. Spread one leg of the cotter pin a little so it cannot fall out. If you bend the legs back too much you weaken them.

13. Replace the hub cap.

BRASS DRIFT OR SOFT BAR

OUTER BEARING CONE AND ROLLER

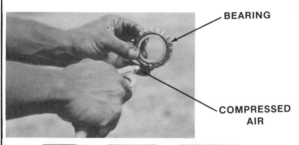

BEARING

COMPRESSED AIR

HUB CAP END
INNER CONE

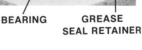

BEARING   GREASE SEAL RETAINER

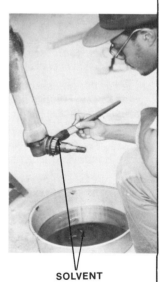

SOLVENT

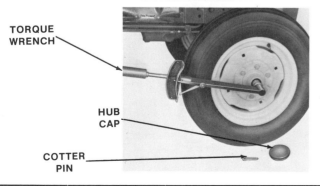

TORQUE WRENCH

HUB CAP

COTTER PIN

BRAKE PEDAL FREE TRAVEL (for mechanical, nonpower brakes)

Most tractors, and some other machines, have two brake pedals. If you push both brake pedals, you brake both drive wheels. If you push one pedal, you only brake one drive wheel.

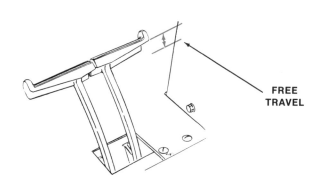

For safe, even stops both brakes must start braking at the same time or the machine will pull to one side. However, brake pedals normally move down a little before they start applying the brakes. This movement is called free travel and it must be adjusted so it is the same on both pedals. Here's how:

1. First, measure the free travel by pulling each pedal down by hand until it stops, and measure this distance.

2. If it isn't the same, adjust the linkage at the brake pedal or brake housing. The operator's manual will tell you how much free travel the brake pedals should have.

3. On tractors with interlocking brake pedals: adjust the pedal with the interlock first, then adjust the other.

LUBRICATING BEARINGS

Some bearings are filled with grease and sealed. Others need grease every so many hours. Your operator's manual will tell you when and where to grease bearings. Greasing at the recommended time makes machines last longer. When you grease bearings:

● *Always use the recommended lubricant. High quality multipurpose grease is good for most agricultural use.*

● *Store lubricants in clean, air tight containers, and do not let water or dirt get in.*

● *Always wipe grease fittings clean before attaching a grease gun so you don't force dirt in with the grease. After lubricating, wipe off excess grease so it can't collect dirt.*

● *Press the grease gun straight onto the grease fitting and pump slowly. Remove the gun by swinging it slowly to the side and pulling back.*

● *Don't lubricate more often than necessary or use too much grease. Excess grease leaks on other parts like drive belts, brake linings, and clutches and makes them slip.*

● *Be careful with a high pressure grease gun. High pressure can rupture grease seals.*

● *Grease when bearings are warm. The grease will flow better.*

Make it a habit to grease your machine at the recommended times. The machine will work better and last longer. While greasing, look for damaged parts, missing bolts, and other problems.

V-BELTS

Proper belt tension is important. A loose belt will slip, grab, and even break. A tight belt will overheat, stretch, and damage bearings, pulleys, and shafts.

Remember, V-belts should ride on the sides of pulleys, not on the bottom of the pulley grooves. Your operator's manual will state the amount of tension each belt should have. Usually belts should deflect about 15 - 20 mm (9/16 - 3/4 inch) between pulleys.

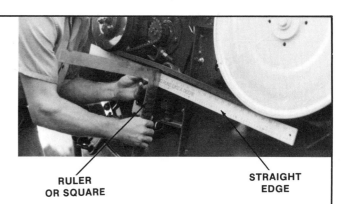

RULER
OR SQUARE

STRAIGHT
EDGE

 CAUTION: Never attempt to check or adjust belts while they are moving. You could be injured.

Check the tension as follows:

1. **Lay a straightedge across the pulleys.**

2. **Push the belt at a place about halfway between the pulleys.**

3. **Measure with a ruler how far you pushed the belt.**

You can also measure belt tension by pulling on the belt with a spring scale and measuring the deflection at the recommended number of kgs (pounds) of pull. The operator's manual will tell you the kgs (pounds).

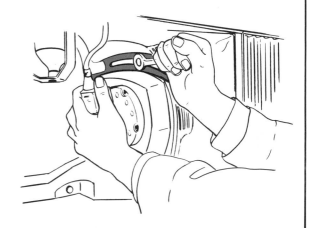

If the belt tension is wrong:

1. **Loosen the belt tightener (bolts and bracket assembly that hold the belt-driven part in position).**

2. **Move the tightener so tension is correct.**

3. **Retighten the belt tightener.**

PULLEY ALIGNMENT

Pulleys must be in line with each other. Use a straightedge to check. The straightedge must touch the pulleys at four places as shown.

If the pulleys are out of alignment, move the pulleys in or out on their shafts so they are aligned. If the shafts or pulleys are bent, replace them.

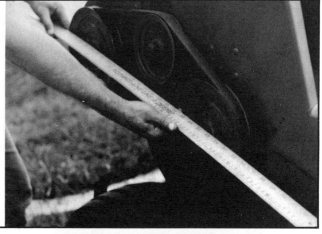

REMOVING AND INSTALLING BELTS

WRONG!
NEVER FORCE
A BELT

If you take off a V-belt, loosen the belt tightener first so the belt can slip off easily. **Never force** a belt on or off a pulley. Forcing will break the cords inside. If several belts are combined to drive a part and one or two are damaged, replace them all. Otherwise, the load will be uneven and the belts will wear out fast. Always replace combination belts in matched sets to be sure they are exactly the same size.

DAMAGED BELTS

RECOGNIZING THE CAUSES OF BELT FAILURES WILL HELP YOU AVOID PROBLEMS. EXAMINE A BROKEN BELT CAREFULLY. TRY TO FIGURE OUT WHAT BROKE IT. IF YOU KNOW WHAT BROKE IT, YOU MAY BE ABLE TO PREVENT ANOTHER BREAK.

INSPECT ALL BELTS ON YOUR MACHINE REGULARLY AND LOOK FOR SIGNS OF DAMAGE AND WEAR. IF BELTS LOOK LIKE THEY ARE GOING TO FAIL, REPLACE THEM.

1. BASE CRACKING

4. SLIP BURN

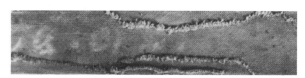

2. FABRIC RUPTURE

5. GOUGED EDGE

3. COVER TEAR

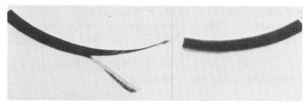

6. RUPTURED CORDS

7. WORN SIDES

TROUBLE SHOOTING OF V-BELT WEAR

SYMPTOM	CAUSE	REMEDY
1. BASE CRACKING	Normal aging. Weather rotted inner fabric.	Replace belt.
2. FABRIC RUPTURE	Prying belt onto sheave. Worn sheaves. Belt too tight.	Loosen belt tighteners before installing belt. Do not tighten belt too much. Also, replace belt.
3. COVER TEAR	Belt coming into contact with some part on machine.	Find interference and eliminate it. Replace belt if cover tear is deep.
4. SLIP BURN	Belt operated too loose. When operated under load, it finally grabbed and snapped.	Tighten belt properly. Also, replace belt.
5. GOUGED EDGE	Damaged sheave or interference from some part on machine.	Check condition of sheaves and check for interference. Also, replace belt.
6. RUPTURED CORDS	Driven sheave locked and drive sheave burned area of belt because belt would not rotate with sheave.	Avoid overloading drive and lubricate bearings to prevent bearing seizure. Also, replace belt.
7. WORN SIDES	Long operation without enough tension.	Check belt tension regularly and keep it properly tightened. Also, replace belt.

116

CHAIN DRIVES

Chains should have enough slack to flex a little between sprockets. Chains that are too tight wear out fast.

Loose chains vibrate and whip. Keep chains adjusted so they can move about 2 percent of their center length.

 CAUTION: Never touch a moving chain.

Chain sprockets, like belt pulleys, can get out of alignment. Use a straightedge to check their alignment, just as you would with belts and pulleys.

Lubricating a chain properly can increase its life five or six times. Dripping oil on the outside of a chain will not lubricate it. The oil cannot get to the bearing surfaces between the links. To lubricate a chain properly, you must get the oil down between the links, where it can do some good.

Oiling the chain while it is still warm and then running it slowly for a while will help the oil get down between the links. Most operator's manuals recommend oiling chains every night.

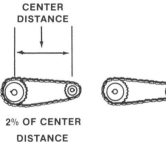

CENTER DISTANCE

2% OF CENTER DISTANCE

CORRECT SLACK TOO TIGHT TOO LOOSE

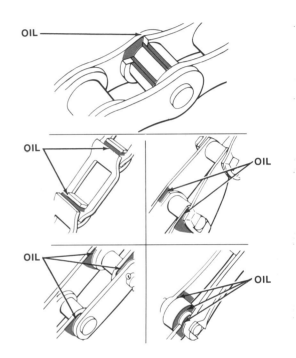

OPEN GEAR DRIVES

 CAUTION: Do not lubricate open gear drives while machine is running. Shut off the engine and wait for all moving parts to stop before you lubricate gears.

Open gears may be lubricated by regularly brushing or squirting oil on them. Or, they may be automatically lubricated with a drip oiler.

Always use the kind of oil recommended in your operator's manual. If oil is too thin, it can't protect the gears. If it is too thick, it can't flow down between the gears to lubricate them.

Dirt is a serious problem with open gears. Dirt between gear teeth grinds away the metal. Flush the gears with plenty of oil to float off the dirt, or clean them with a safe solvent or diesel fuel before lubricating them.

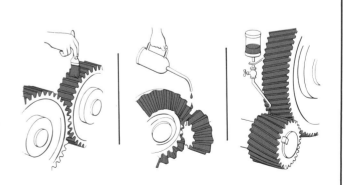

CLEANING YOUR MACHINE

⚠️ **CAUTION: Never use gasoline (petrol) to clean machine parts. It is a serious fire hazard.**

Wash off chemicals, grease, and dirt. Clean engines, crankcases, and transmissions radiate heat better than dirty ones. They stay cooler, and that reduces wear and makes oil last longer. Also, you can see damage and leaks when parts are clean.

Your own safety is another reason for washing. Slick steps created by grease and mud, can cause a serious fall.

So, clean machine parts when they are warm. If they are too hot, the solvent will evaporate. If too cold, grease can't be loosened.

Wear safety glasses if you use a high-pressure hose.

Clean machines with diesel fuel or a safe solvent recommended in the owner's manual. Never clean with gasoline (petrol).

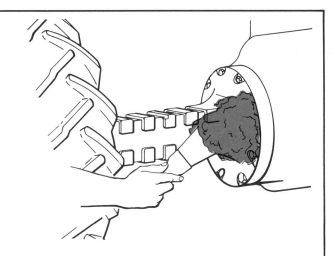

SUMMARY

One of the easiest and most effective ways you can make tires last longer is by keeping the right pressure in them.

Another thing you can do is make sure you have the right ballast for good traction without overloading the machine.

Check the oil level in the steering gear box, on tractors with manual steering, and keep the steering adjusted.

Measure the toe-in and adjust it too.

Remove, clean, and pack the wheel bearings. Be sure you tighten and pin the slotted nut.

Check and adjust the free travel in the brake pedals. Both pedals should start brake action at the same time.

Inspect V-belts, pulleys, and chain drives. Do not let them get too loose, too tight, or out of alignment.

Lubricate and clean the dirt off open gears often.

Clean grease fittings and lubricate them as recommended in the operator's manual.

Wash grease and dirt off your machine often. Keep it clean so it will run cooler, be safer, and so you can see damage.

DO YOU REMEMBER?

1. Explain the tire problems caused by:

a. Overinflation

b. Underinflation

2. Why should tire pressure be measured when tires are cold?

3. Describe wheel tracks when wheels:

a. Have proper weight

b. Have too much weight

c. Have too little weight

4. If liquid ballast is used, tires must be completely filled. True   False

5. When measuring toe-in, should the front edge of wheels on the steering axle be closer or wider than the rear edge?

6. To check brake action, spin wheels rapidly and stop them suddenly.   True   False

7. What happens if V-belts are too loose?

SAFETY
CHAPTER 11

INTRODUCTION

Each year nearly 2000 people die in machinery related accidents in North America; many more around the world. In addition many thousands receive serious injuries that will disable them for the rest of their lives.

Stop....Think....

You can quickly become one of those statistics, if you don't follow safe operating and maintenance practices.

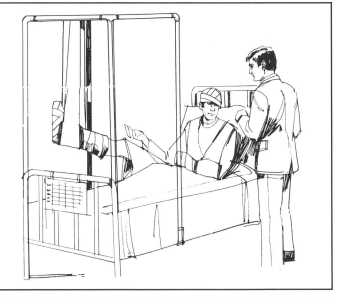

FOLLOW ALL SAFETY INSTRUCTIONS

Stop and think about the maintenance procedures. What could happen? How could it happen? What is the safest way? Read all safety instructions and follow them every time.

GET FAMILIAR WITH EACH MACHINE

Don't rush into a maintenance job. Get very familiar with the specific functions and potential safety hazards of each machine. Always refer to the operator's manual of each machine. Even similar machines do not always have identical systems. So be very careful. Follow all safety instructions...every time. Remember: You are protecting yourself and others by being careful.

RECOGNIZE SAFETY INFORMATION

This is the safety-alert symbol. When you see this symbol on your machine or in this manual, be alert to the potential for personal injury.

Follow recommended precautions and safe operating practices.

UNDERSTAND SIGNAL WORDS

A signal word—DANGER, WARNING, or CAUTION— is used with the safety-alert symbol. DANGER identifies the most serious hazards.

DANGER or WARNING safety signs are located near specific hazards. General precautions are listed on CAUTION safety signs. CAUTION also calls attention to safety messages in this manual.

FOLLOW SAFETY INSTRUCTIONS

Carefully read all safety messages in this manual and on your machine safety signs. Keep safety signs in good condition. Replace missing or damaged safety signs. Be sure new equipment components and repair parts include the current safety signs. Replacement safety signs are available from your dealer.

Learn how to operate the machine and how to use controls properly. Do not let anyone operate without instruction.

Keep your machine in proper working condition. Unauthorized modifications to the machine may impair the function and/or safety and affect machine life.

If you do not understand any part of this manual and need assistance, contact your dealer.

PRACTICE SAFE MAINTENANCE

Understand service procedure before doing work. Keep area clean and dry.

Never lubricate or service machine while it is moving. Keep hands, feet and clothing from power-driven parts. Disengage all power and operate the controls to relieve pressure. Lower equipment to the ground. Stop the engine. Remove the key. Allow machine to cool.

Securely support any machine elements that must be raised for service work.

Keep all parts in good condition and properly installed. Fix damage immediately. Replace worn or broken parts. Remove any buildup of grease, oil or debris.

Disconnect battery ground cable(s) (−) before making adjustments on electrical system or welding on machine.

REMEMBER: Always turn off the engine before you work on a machine.

SERVICE MACHINE SAFELY

Tie long hair behind your head. Do not wear a necktie, scarf, loose clothing or necklace when you work near machine tools, or moving parts. If these items were to get caught, severe injury could result.

Remove rings and other jewelry to prevent electrical shorts and entanglement in moving parts.

SUPPORT MACHINE PROPERLY

Always lower the attachment or implement to the ground before you work on the machine. If you must work on a lifted machine or attachment, securely support the machine or attachment.

Do not support the machine on cinder blocks, hollow tiles or props that may crumble under continuous load. Do not work under a machine that is supported solely by a jack. Follow recommended procedures in this manual.

SERVICE FRONT-WHEEL-DRIVE TRACTOR SAFELY

When servicing front-wheel-drive tractor with the rear wheels supported off the ground and rotating wheels by engine power, always support front wheels in a similar manner. Loss of electrical power or transmission/hydraulic system pressure will engage the front driving wheels, pulling the rear wheels off the support if front wheels are not raised. Under these conditions, front drive wheels can engage even with switch in disengaged position.

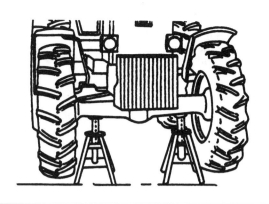

RETIGHTEN WHEEL NUTS

Retighten machine wheel nuts at the intervals specified in the machine operators manual.

KEEP CAB/ROPS INSTALLED PROPERLY

Make certain all parts are reinstalled correctly if the cab or roll-over protective structure (ROPS) is loosened or removed for any reason. Tighten mounting bolts to specified torque.

Protection offered by cab or ROPS will be impaired if subjected to structural damage, involved in an overturn incident or altered in any way by welding, bending, drilling or cutting. A damaged cab or ROPS should be replaced, not reused.

HANDLE FUEL SAFELY—AVOID FIRES

Handle fuel with care: it is highly flammable. Do not refuel the machine while smoking or when near open flame or sparks.

Always stop engine before refueling machine. Fill fuel tank outdoors.

Prevent fires by keeping machine clean of accumulated trash, grease, and debris. Always clean up spilled fuel.

AVOID BEING CRUSHED

There are many crush points on a machine. A crushing injury or death is caused, if you get between two parts that move towards each other, like the front and rear portions of a four-wheel-drive articulated tractor.

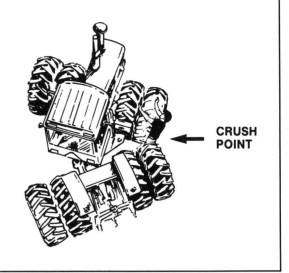

CRUSH POINT

You can also get crushed, if you get between a stationary part and a moving part, like between the 3-point hitch and the tractor. Another way to get crushed is to be under a machine that is not well blocked and the machine falls on you.

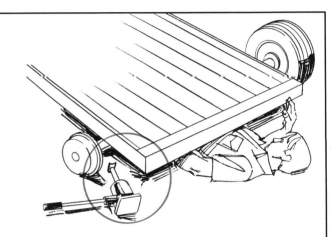

WORKING UNDER HEAVY OBJECTS THAT AREN'T SECURELY BLOCKED MAY RESULT IN CRUSHING INJURIES

BE AWARE OF STORED ENERGY

Stored energy can work for you, or it can be carelessly released and cause injury.

In many farm machines, energy is stored so it can be released at the right time, in the right way for you. Here are some components that store energy. You should recognize them as you use and service farm machinery.

- **Springs**
- **Hydraulic Systems**
- **Compressed Air**
- **Electricity**
- **Raised Loads**
- **Loaded Mechanisms**

Always release or disconnect the stored energy before you begin maintenance work.

SPRING UNDER COMPRESSION

PARK MACHINE SAFELY

Before working on the machine:

- Lower all equipment to the ground.
- Stop the engine and remove the key.
- Disconnect the battery ground strap.
- Hang a "DO NOT OPERATE" tag in operator station.

ILLUMINATE WORK AREA SAFELY

Illuminate your work area adequately but safely. Use a portable safety light for working inside or under the machine. Make sure the bulb is enclosed by a wire cage. The hot filament of an accidentally broken bulb can ignite spilled fuel or oil.

USE SAFETY LIGHTS AND DEVICES

Slow moving tractors, self-propelled equipment and towed implements or attachments can create a hazard when driven on public roads. They are difficult to see, especially at night. Avoid personal injury or death resulting from collision with a vehicle.

Flashing warning lights and turn signals are recommended whenever driving on public roads. To increase visibility, use the lights and devices provided with your machine. For some equipment, install extra flashing warning lights.

Keep safety items in good condition. Replace missing or damaged items. An implement safety lighting kit is available from your dealer.

OBSERVE ROAD TRAFFIC REGULATIONS

Always observe local road traffic regulations when using public roads.

Replace the Slow-Moving-Vehicle Emblem (SMV) if it is missing.

PREVENT MACHINE RUNAWAY

Avoid possible injury or death from a machine runaway.

Do not start the engine by shorting across starter terminals. Machine will start in gear if normal circuitry is bypassed.

NEVER start engine while standing on ground. Start engine only from operator's seat, with the transmission in neutral or "Park".

PREPARE FOR EMERGENCIES

Be prepared if a fire starts.

Keep a first aid kit and fire extinguisher handy.

Keep emergency numbers for doctors, ambulance service, hospital and fire department near your telephone.

WEAR PROTECTIVE CLOTHING

Wear close fitting clothing and safety equipment appropriate to the job.

Prolonged exposure to loud noise can cause impairment or loss of hearing.

Wear a suitable hearing protective device such as earmuffs or earplugs to protect against objectionable or uncomfortable loud noises.

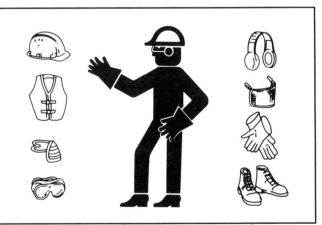

DISCONNECT ELECTRICAL CIRCUIT

Disconnect battery ground strap(s) before carrying out any electrical repairs or welding on the machine.

HANDLE BATTERIES SAFELY

Sulfuric acid in battery electrolyte is poisonous. It is strong enough to burn skin, eat holes in clothing and cause blindness if splashed into eyes.

Avoid the hazard by:
1. Filling the batteries in a well-ventilated area.
2. Wearing eye protection and rubber gloves.
3. Avoiding breathing in the fumes when electrolyte is added.
4. Avoiding spilling or dripping electolyte.
5. Using the proper jump start procedure.

If you spill acid on yourself:
1. Flush your skin with water.
2. Apply baking soda or lime to help neutralize the acid.
3. Flush your eyes with water for 10 - 15 minutes.
Get medical attention immediately.

If acid is swallowed:
1. Drink large amounts of water or milk.
2. Then drink milk of magnesia, beaten eggs or vegetable oil.
3. Get medical attention immediately.

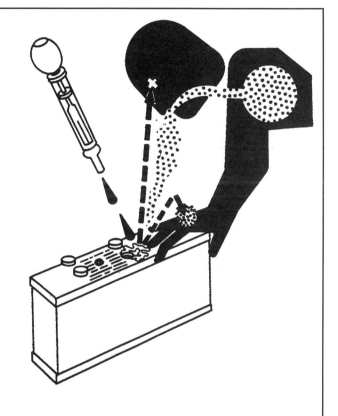

STORE ATTACHMENTS SAFELY

Store attachments such as dual wheels, cage wheels and loaders can fall and cause serious injury or death.

Securely store attachments and implements to prevent falling. Keep playing children and bystanders away from storage area.

WORK IN VENTILATED AREA

Engine exhaust fumes can cause sickness or death. If it is necessary to run an engine in an enclosed area, remove the exhaust fumes from the area with an exhaust pipe extension.

If you do not have an exhaust pipe extension, open the doors and get outside air into the area.

HANDLE STARTING FLUID SAFELY

Starting fluid is highly flammable.

Keep all sparks and flames away when using it. Keep starting fluid away from batteries and cables.

To prevent accidental discharge when storing the pressurized can, always keep the cap on the container and store in a cool, protected location.

Do not incinerate or puncture starting fluid containers, even when empty.

SERVICE COOLING SYSTEM SAFELY

Danger of scalding!

With the engine shut off, first loosen the radiator cap or expansion tank cap only to the first stop to relieve the pressure before removing the cap completely.

Top off cooling system with coolant only when engine is shut off.

AVOID HIGH-PRESSURE FLUIDS

The hydraulic system and the fuel injection system contain fluids under very high pressure. Small leaks in these systems will allow oil or fuel to escape with high velocity.

Escaping fluid under pressure can penetrate the skin causing serious injury.

Avoid the hazard by relieving pressure before disconnecting hydraulic or other lines. Tighten all connections before applying pressure.

Search for leaks with a piece of cardboard. Protect hands and body from high pressure fluids.

If an accident occurs, see a doctor immediately. Any fluid injected into the skin must be surgically removed within a few hours or gangrene may result. Doctors unfamiliar with this type of injury should contact other knowledgeable medical sources.

OBSERVE ENVIRONMENTAL PROTECTION REGULATIONS

Be mindful of the environment and ecology.

Before draining any fluids, find out the correct way of disposing of them.

Observe the relevant environmental protection regulations when disposing of oil, fuel, coolant, brake fluid, filters and batteries.

AVOID HARMFUL ASBESTOS DUST

Avoid breathing dust that may be generated when handling components containing asbestos fibers. Inhaled asbestos fibers may cause lung cancer.

Components that may contain asbestos fibers are brake pads, brake band and lining assemblies, clutch plates and some gaskets. The asbestos used in these components is usually found in a resin or sealed in some way. Normal handling is not hazardous as long as airborne dust containing asbestos is not generated.

Avoid creating dust. Never use compressed air for cleaning. Avoid brushing or grinding of asbestos-containing materials. When servicing, wear an approved respirator. A special vacuum cleaner is recommended to clean asbestos. If not available, wet the asbestos-containing materials with a mist of oil or water.

Keep bystanders away from the area.

Please note designations on spare parts.

STAY CLEAR OF ROTATING DRIVELINES

Entanglement in rotating driveline can cause serious injury or death. When operating with the PTO, no-one must be allowed to remain in the vicinity of the rotating PTO stub shaft or drive shaft. Always ensure that the guards for PTO drive shaft, PTO stub shaft and drive shaft guards are in position and that rotating shields turn freely.

Wear close-fitting clothing. Stop the engine and be sure PTO driveline is stopped before making adjustments, connections or cleaning out PTO-driven equipment.

As soon as PTO drive shaft has been removed, reinstall guard over PTO stub shaft.

USE TOOLS THAT FIT

Many serious accidents happen, when the wrong tools are used or when tools are used that don't fit. Make sure you use metric dimensioned tools on metric products, and inch-dimensioned tools on products of customary design. Also, don't use tools for purposes they are not designed for.

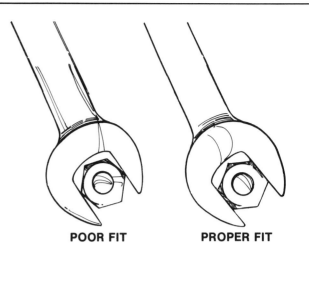

POOR FIT   **PROPER FIT**

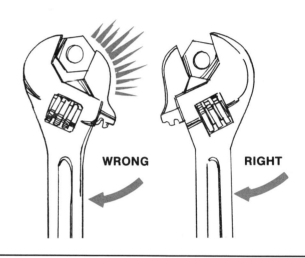

WRONG   **RIGHT**

AVOID HEATING NEAR PRESSURIZED FLUID LINES

Flammable spray can be generated by heating near pressurized fluid lines, resulting in severe burns to yourself and bystanders. Do not heat by welding, soldering, or using a torch near pressurized fluid lines or other flammable materials. Pressurized lines can be accidentally cut when heat goes beyond the immediate flame area.

PROTECT YOUR EYES

Eyesight is one of the most precious abilities we have. So protect your eyes very carefully. Wear safety glasses or face shields, when you are using a chisel or punch. Wear a welding shield or a helmet when you are welding. Keep shield in place when you are chipping slag.

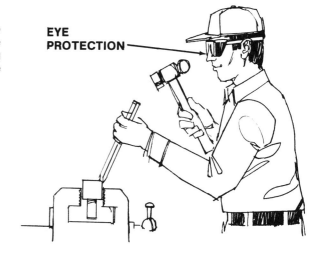

EYE PROTECTION

WEAR GOGGLES UNDER HELMET

USE CLEAR LENS IN WELDING HELMET

PROTECT YOUR HEARING

Machines that are running make a lot of noise. Prolonged exposure to noise will damage your ability to hear. Therefore you should always wear hearing protection such as ear plugs or ear muffs.

SYMBOLS THAT COMMUNICATE

Everywhere you look there are man-made symbols that communicate messages. Do you recognize all of the symbols shown? Do you know what they mean? Try to identify each one to see if you understand it.

1. *Compulsory way for pedestrians.*

2. *Dangerous curve.*

3. *Beware of animals.*

4. *No U-turns.*

5. *No stopping.*

6. *Speed limits for light and heavy motor vehicles.*

7. *No passing.*

8. *Road narrows.*

9. *Slippery road.*

10. *Compulsory minimum speed.*

SUMMARY

Safety is the most important consideration when you perform maintenance on a machine. A very small careless act can result in a major injury. . . . possibly death. Whatever you do. . . . safety comes first.

DO YOU REMEMBER?

How many people, like yourself, die as a result of machinery accidents each year?

Are you familiar with how crushing injuries occur?

What must you do when you suspect that high pressure liquids have penetrated your skin?

Are cement or cinder blocks suitable for supporting a heavy tractor?

SYMBOLS THAT COMMUNICATE

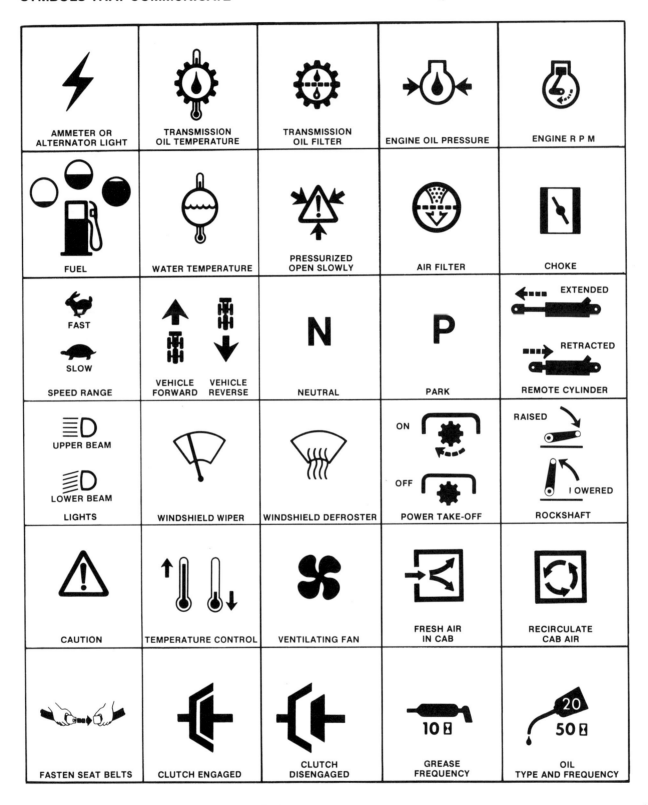

AMMETER OR ALTERNATOR LIGHT	TRANSMISSION OIL TEMPERATURE	TRANSMISSION OIL FILTER	ENGINE OIL PRESSURE	ENGINE R P M
FUEL	WATER TEMPERATURE	PRESSURIZED OPEN SLOWLY	AIR FILTER	CHOKE
SPEED RANGE (FAST / SLOW)	VEHICLE FORWARD / VEHICLE REVERSE	NEUTRAL	PARK	REMOTE CYLINDER (EXTENDED / RETRACTED)
LIGHTS (UPPER BEAM / LOWER BEAM)	WINDSHIELD WIPER	WINDSHIELD DEFROSTER	POWER TAKE-OFF (ON / OFF)	ROCKSHAFT (RAISED / LOWERED)
CAUTION	TEMPERATURE CONTROL	VENTILATING FAN	FRESH AIR IN CAB	RECIRCULATE CAB AIR
FASTEN SEAT BELTS	CLUTCH ENGAGED	CLUTCH DISENGAGED	GREASE FREQUENCY	OIL TYPE AND FREQUENCY

Courtesy of the American Society of Agricultural Engineers.
For a complete listing of symbols see ASAE Standard S304.5

APPENDICES

OPERATIONAL CHECKOUT

Several machinery manufacturers provide a new approach to finding the cause of the most common malfunctions without the use of complicated diagnostic gauges and instruments. These operational check-out procedures are based on the maintenance person's ability to Look, Listen and Feel. (Do you remember what was said on page 12? Please turn to it again.)

On the following two pages you will find samples of operational check-outs. Of course, each machine will require it's very own specific procedure.

Use this procedure to check all systems and functions on the machine. It is designed so you can make a quick check of the operation of the machine while doing a walk around inspection and performing specific checks from the operator's seat.

Should you experience a problem with your machine, you will find helpful diagnostic information in this checkout that will pinpoint the cause. This information may allow you to perform a simple adjustment which will reduce the downtime of your machine. Use the table of contents to help find adjustment procedures.

The information you provide after completing the operational checkout will allow you or your dealer to pinpoint the specific test or repair work needed to restore the machine to design specifications.

A location will be required which is level and has adequate space to complete the checks. No tools or equipment are needed to perform the checkout.

Complete the necessary visual checks (oil levels, oil condition, external leaks, loose hardware, linkage, wiring, etc.) prior to doing the checkout. The machine must be at operating temperature for many of the checks.

Start at the top of the left column and read completely down column before performing check, follow this sequence from left to right. In the far right column, if no problem is found (OK), you will be instructed to go to next check. If a problem is indicated (NOT OK), you will be referred to either a section in this manual or to your dealer.

TURN SIGNAL AND FLASHER CHECK	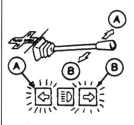 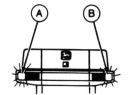	Move turn signal lever to left turn position (A) and then right turn position (B). *LOOK: Left front and rear amber lights and dash indicator (A), then right front and rear amber lights and dash indicator (B) must be flashing.*	**OK:** Go to next check **NOT OK:** If turn signals do not work, check turn signal fuse. **IF OK:** Go to your dealer.
WARNING FLASHER CHECK	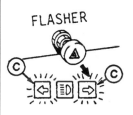 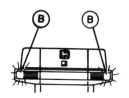	Pull out warning flasher switch knob. *LOOK: Front (A) and rear (B) amber lights and both dash indicators (C) must be flashing.*	**OK:** Go to next check **NOT OK:** Check turn signal fuse. **IF OK:** Go to your dealer.
BRAKE LIGHT CHECK	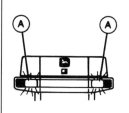	Depress brake pedal and observe brake lights (A). *LOOK: Brake lights must come on.*	**OK:** Go to next check **NOT OK:** Check brake light fuse. **IF OK:** Go to your dealer.
HORN CIRCUIT CHECK Key ON.		Push button. *LISTEN: Horn (A) must sound.*	**OK:** Go to next check **NOT OK:** Check fuse number 5. **IF OK:** Go to your dealer.

MECHANICAL FRONT WHEEL DRIVE (MFWD) DRIVING CHECKS

MFWD LIMITED-SLIP DIFFERENTIAL AND CONTROL LINKAGE CHECK	Shift transaxle to 1st reverse. Engage MFWD and drive machine. Run engine at approximately 1200 rpm. Turn steering wheel for a full left or right turn and observe how the front tires attempt to slide sideways (side load), and the amount of tire scuffing. Disengage MFWD. *LOOK: Tire side loading and scuffing must stop when MFWD is disengaged.* *NOTE: If tires attempt to slide sideways and tire scuffing is seen with MFWD engaged, limited-slip differential is working and power is being transmitted to MFWD.*	**OK:** Go to next check. **NOT OK:** Move MFWD control linkage to feel engagement detents. If no detents, inspect linkage to transfer case. If OK, go to Troubleshooting chapter, MFWD, No Power To MFWD. Go to your dealer.
ENGINE AND TORQUE CONVERTER CHECK	With loader bucket level, and cutting edge at the centerline of front wheels, position machine against a dirt bank or immovable object. Engage MFWD and differential lock. Shift transaxle to 1st forward. Increase engine speed to fast idle. *LOOK: All four wheels must turn.*	**OK:** Go to next check. **NOT OK:** If all wheels stop, a torque converter problem is indicated. If the front wheels stop, a MFWD problem is indicated. Go to Troubleshooting chapter, Power Train, Machine Lacks Power Or Moves Slow. Go to your dealer.
MFWD GEAR AND PINION CHECK	Drive machine at transport speed with MFWD engaged, then disengaged. *LISTEN: MFWD MUST NOT whine.*	**OK:** Go to next check. **NOT OK:** If MFWD whines, check oil levels and fill to correct levels. If OK, check backlash. Go to your dealer.

TROUBLESHOOTING CHART

ENGINE

Problem	Cause	Remedy
1. ENGINE HARD TO START OR WILL NOT START	No fuel or improper fuel.	Fill tank. If wrong fuel, drain and refill with proper fuel.
	Water or dirt in fuel or dirty filters.	Check out fuel supply. Replace or clean filters.
	Air in fuel system (Diesel).	Bleed air from system.
	Low cranking speed.	Charge or replace battery, or service starter as necessary.
	Improper timing.	Check distributor (Spark-Ignition).
		Have service shop check injection pump (Diesel).
	Defective coil or condenser (Spark-Ignition).	Replace coil or condenser.
	Pitted or burned distributor points (Spark-Ignition).	Clean or replace points.
	Cracked distributor cap or eroded rotor (Spark-Ignition).	Replace cap or rotor.
	Distributor wires loose or installed in wrong order (Spark-Ignition).	Push wire into sockets. Install wires in correct firing order.
	Fouled or defective spark plugs (Spark-Ignition).	Clean and regap plugs or replace them.
	Poor injection nozzle operation (Diesel).	Have service shop clean or repair nozzles.
	Liquid fuel in lines (LP-Gas).	Always turn on vapor valve when starting the engine.
	Engine flooded.	Wait several minutes before attempting to start engine. Do not choke engine again.
2. ENGINE STARTS BUT WILL NOT RUN PROPERLY	Fuel problem — dirt, air restrictions, or clogged filters.	Check fuel supply, bleed system (Diesel), check for line restrictions, and clean or replace fuel and air filters and screens.
	Carburetor needs adjustment (Spark-Ignition).	Adjust idle and load air-fuel mixtures.
	Defective coil or condenser (Spark-Ignition).	Replace as necessary.
	Defective ignition resistor or key switch (Spark-Ignition).	Replace resistor or switch.
	Pitted or burned distributor points (Spark-Ignition).	Clean or replace points.
	Fouled or defective spark plugs (Spark-Ignition).	Clean and regap plugs or replace them.
	Cracked distributor cap or eroded rotor (Spark-Ignition).	Replace cap or rotor.
3. ENGINE DETONATES (Gasoline [Petro])	Wrong type of fuel.	Use proper octane fuel.

(continued on next page)

Problem	Cause	Remedy
4. ENGINE PRE-IGNITES (Gasoline [Petro])	Distributor timed too early.	Retime distributor.
	Distributor advance mechanism stuck.	Free mechanism.
	Faulty spark plugs or spark plug heat range too high.	Install new plugs that have proper heat range.
5. ENGINE BACK FIRES — (Spark-Ignition)	Spark plug cables installed wrong.	Install in correct firing order.
	Carburetor mixture too lean.	Adjust carburetor.
6. ENGINE KNOCKS	Improper distributor/timing (Spark-Ignition).	Time distributor.
	Improper injection pump timing (Diesel).	Have service shop check Injection pump.
	Worn engine bearings or bushings.	Have service shop replace.
	Loose bearing caps.	Have service shop tighten to proper torque.
	Foreign matter in the cylinder.	Have service shop remove material and repair engine.
7. ENGINE OVERHEATS	Defective radiator cap.	Replace cap.
	Radiator core plugged with dirt and debris.	Clean radiator core.
	Defective thermostat.	Replace thermostat.
	Loss of coolant.	Check for leaks and correct.
	Loose fan belt.	Adjust tension.
	Cooling system has scale deposit build-up.	Use cooling system cleaner to remove scale.
	Overloaded engine.	Reduce load or shift into a lower gear.
	Incorrect engine timing.	Time distributor (Spark-Ignition).
		Have service shop time injection pump (Diesel).
	Engine low on oil.	Add oil to the proper level.
	Wrong type of fuel.	Use recommended fuel.
8. LACK OF POWER	Air cleaner dirty or obstructed.	Clean or replace air cleaner. Remove obstruction.
	Restriction in fuel lines, filters or carburetor.	Clean plugged parts.
	Wrong type of fuel.	Use recommended type.
	Frost at fuel-lock strainer (LP-Gas).	Clean strainer.
	Governor not operating properly.	Have service shop adjust or repair.

(continued on next page)

Problem	Cause	Remedy
	Valves in engine head leaking.	Have service shop recondition.
	Incorrect valve clearance.	Adjust clearance.
	Low engine compression.	Have service shop check and repair engine.
	Incorrect timing.	Time distributor (Spark-Ignition).
		Have service shop time injection pump (Diesel).
	Carburetor improperly adjusted or dirty (Spark-Ignition).	Have service shop clean and adjust carburetor.
	Wrong spark plugs or plugs fouled (Spark-Ignition).	Clean, regap or replace spark plugs.
	Distributor points burned (Spark-Ignition).	Replace points and condenser.
	High engine operating temperature.	See "ENGINE OVERHEATS" above.
	Low engine operating temperature.	Check thermostat.
	Vent on fuel tank plugged (Gasoline [Petrol] and Diesel).	Check fuel tank cap.
9. ENGINE USES TOO MUCH OIL	Crankcase oil too light.	Use recommended weight of oil.
	Worn pistons and rings.	Have service shop recondition.
	Worn valve guides or stem oil seals.	Have service shop replace.
	External oil leaks.	Eliminate leaks.
	Oil pressure too high.	Have service shop adjust pressure.
	Restricted air intake system.	Check system and relieve restriction.
10. OIL PRESSURE TOO LOW	See "ENGINE USES TOO MUCH OIL" above.	
11. ENGINE USES TOO MUCH FUEL	Clogged or dirty air cleaner.	Clean or replace air cleaner.
	Improper type fuel used.	Drain fuel tank and fill with recommended fuel.
	Engine overloaded.	Reduce load or shift into lower gear.
	Improper valve clearance.	Adjust valve clearance.
	Engine out of time.	Time distributor (Spark-Ignition).
		Have service shop time injection pump (Diesel).
	Incorrect carburetor adjustment.	Adjust carburetor.
	Engine not operating at proper temperature.	Check thermostat.
	Choke in closed position.	Open choke or adjust linkage if necessary.
12. ENGINE EXHAUSTS BLACK OR GRAY SMOKE	Improper type of fuel.	Drain fuel tank and fill with recommended fuel.
	Clogged or dirty air cleaner.	Clean or replace air cleaner.
	Defective muffler.	Replace muffler.
	Engine overloaded.	Reduce load or shift into a lower gear.
	Fuel injection system faulty.	Have service shop determine cause and repair.

(continued on next page)

ENGINE — CONTINUED

Problem	Cause	Remedy
	Engine out of time.	Time distributor (Spark-Ignition).
		Have service shop time injection pump (Diesel).
	Incorrect carburetor adjustment.	Adjust carburetor.
13. ENGINE EXHAUSTS WHITE SMOKE	Improper type fuel.	Drain fuel tank and fill with recommended fuel.
	Low engine temperature.	Allow engine to warm up to normal temperatures before operating under load.
	Defective thermostat.	Replace thermostat.
	Engine out of time.	Time distributor (Spark-Ignition).
		Have service shop time injection pump (Diesel).

ELECTRICAL SYSTEMS

Problem	Cause	Remedy
1. LOW BATTERY OUTPUT	Low electrolyte level.	Add distilled water to proper level.
	Low specific gravity.	See "LOW BATTERY CHARGE" below.
	Defective battery cell.	Replace battery.
	Cracked or broken case.	Replace battery.
	Low battery capacity.	Replace battery with one of recommended capacity.
2. BATTERY USES TOO MUCH WATER	Cracked battery case.	Replace battery.
	Overcharged battery.	Apply load to battery and have voltage regulator checked.
3. LOW BATTERY CHARGE	Excessive loads from added accessories.	Remove excessive loads or install larger alternator.
	Excessive engine idling.	Allow engine to idle only when necessary.
	Continuous drain on battery.	Clean battery top.
		Check for component grounded or shorted.
	High resistance in circuit.	Clean and tighten connections. Replace faulty wiring.
	Faulty charging operation.	See "LOW CHARGING CIRCUIT OUTPUT."
4. LOW CHARGING CIRCUIT OUTPUT	Slipping drive belts.	Adjust belt tension.
	Excessively worn or sticking brushes in alternator or generator.	Replace brushes.
	Defective alternator or generator.	Replace unit or have service shop repair.
5. NOISY GENERATOR OR ALTERNATOR	Defective or badly worn belt.	Replace belt.
	Generator brushes not seated.	Seat brushes.
	Generator commutator worn too much.	Have service shop recondition.

(continued on next page)

142

Problem	Cause	Remedy
	Worn or defective bearings.	Have service shop replace bearings.
	Loose mounting or loose pulley.	Tighten mounting and pulley.
	Misaligned drive belt and pulley.	Realign belt and pulley.
6. SLUGGISH STARTING MOTOR OPERATION	Low battery charge.	Charge battery
	High resistance in circuit.	Clean and tighten wiring connections.
	Defective starting motor.	Have service shop repair.
	Starting motor bearings dry.	Lubricate bearings or replace sealed bearings.
	Engine oil viscosity too high.	Drain oil and replace with viscosity of oil recommended for cold temperatures.
7. STARTING MOTOR WILL NOT OPERATE	Low battery charge.	Charge battery.
	High resistance in circuit.	Clean and tighten connections.
	Starter safety switch open.	Move shift lever to neutral or park position.
	Defective or improperly adjusted starter safety switch.	Have service shop adjust or replace switch.
	Defective starter switch.	Replace switch.
	Defective starter.	Have service shop check, repair or replace starter.
8. MISFIRING OF ENGINE	Improper spark plug heat range.	Replace plugs with hotter or colder range plugs as required.
	Bad plug wiring.	Replace plug wires.
	Worn spark plug electrodes or fouled plugs.	Clean plugs and regap. Replace plugs if necessary.
	Defective spark plugs.	Replace plugs.
	Incorrect distributor timing.	Retime distributor.
	Insufficient voltage available to spark plugs.	See "LOW VOLTAGE AT SPARK PLUG," below.
9. LOW VOLTAGE AT SPARK PLUG	Worn or improperly spaced distributor points.	Adjust point gap.
	Dirty, burned, or pitted points.	Clean or replace points and condenser.
	Defective condenser.	Replace condenser.
	Dirt or moisture in distributor cap.	Clean distributor cap.
	Cracked distributor cap.	Replace cap.
	Eroded distributor rotor.	Replace rotor.
	Defective spark plug cables.	Replace cables.
	Defective ignition coil.	Replace coil.
	Loose wire connections.	Tighten all connections.
10. BUILD-UP OF MATERIAL ON DISTRIBUTOR POINTS	Condenser has improper capacity.	Replace with proper condenser.
11. EXCESSIVE WEAR ON DISTRIBUTOR POINTS RUBBING BLOCK	Inadequate lubricant.	Use cam lubricant to prevent wear.
12. BURNED DISTRIBUTOR POINTS	Loose lead wire or high resistance in condenser.	Tighten lead or replace condenser.
	Wrong method of cleaning distributor points.	Use point file and lintless cloth.
	Oil or foreign material on points.	Clean or replace points.

(continued on next page)

ELECTRICAL SYSTEMS — CONTINUED

Problem	Cause	Remedy
13. DIM LIGHTS	High resistance in circuit or poor ground on lights.	Clean and tighten all connections. Replace faulty wiring.
	Low battery charge.	Charge battery.
	Defective light switch.	Replace switch.
14. GENERATOR OR ALTERNATOR INDICATOR LAMP GLOWS INTERMITTENTLY	Excessive resistance in battery lead to unit or regulator.	Clean and tighten all connections. Replace faulty wiring.
	Excessive internal resistance in generator or alternator.	Replace brushes or take to service department.
		Have unit repaired or replaced.
15. OIL PRESSURE INDICATOR LAMP FAILS TO LIGHT	Burned-out bulb.	Replace bulb.
	Open circuit or excessive resistance in wiring.	Clean and tighten all connections. Replace faulty wiring.
	Defective lamp body.	Replace lamp body.
	Faulty oil pressure switch.	Replace switch.
16. OIL PRESSURE LAMP REMAINS ON WITH STARTING SWITCH OFF	Defective lamp body.	Replace lamp body.
	Grounded wire to oil pressure switch.	Repair or replace wiring.
	Faulty oil pressure switch.	Replace switch.

POWER TRAIN (CLUTCH)

Problem	Cause	Remedy
1. CLUTCH SLIPS	Too little clutch pedal free travel.	Adjust clutch pedal free travel.
	Operator riding clutch pedal.	Do not ride clutch pedal.
	Worn clutch disks.	See dealer for repair.
2. CLUTCH GRABS OR CHATTERS		Take the machine to a service shop for repair or adjustment.
3. CLUTCH SQUEAKS	Clutch release bearing dry.	Lubricate bearing.
	Clutch actuating mechanism dry.	Lubricate linkage and shafts.
4. CLUTCH RATTLES AND VIBRATES		Take the machine to a service shop for repair or adjustment.

POWER TRAIN (MECHANICAL TRANSMISSION)

Problem	Cause	Remedy
1. TRANSMISSION NOISY	Transmission oil level low.	Fill transmission with proper lubricant.
	Worn or broken gears.	See dealer for repair.
2. TRANSMISSION HARD TO SHIFT		Take machine to service shop for repair or adjustment.
3. TRANSMISSION STICKS IN GEAR	Clutch not releasing.	Adjust clutch pedal free travel.
	Shift linkage binding.	Free linkage.
	Worn shift linkage.	Have dealer repair.

(continued on next page)

POWER TRAIN (MECHANICAL TRANSMISSION) — Continued

Problem	Cause	Remedy
4. TRANSMISSION SLIPS OUT OF GEAR		Take machine to service shop for repair or adjustment.
5. TRANSMISSION LEAKS OIL	Oil level too high.	Drain to proper level.
	Gaskets damaged or missing.	Have dealer install new gaskets.
	Drain plug loose.	Tighten drain plug.
	Lubricant foaming excessively.	Use recommended lubricant.

POWER TRAIN (HYDRAULIC ASSIST TRANSMISSION)

Problem	Cause	Remedy
1. MACHINE WON'T MOVE	Cold weather starting clutch disengaged.	Engage clutch.
	Park lock engaged.	Release lock.
	Control linkage binding or disconnected.	Free linkage or connect linkage.
	Oil filter plugged.	Replace oil filter.
2. SHIFTS ERRATICALLY	Shift control disconnected or binding.	Free controls or connect disconnected parts.
3. LOW SYSTEM PRESSURE	Plugged oil filter.	Replace filter.
	Low oil level.	Fill to proper level.
4. TRANSMISSION OVERHEATING	Reservoir oil level too low or too high.	Bring oil level to proper level.
	Plugged oil filter.	Replace filter.
	Plugged core in oil cooler.	Clean core.

POWER TRAIN (TORQUE CONVERTER)

Problem	Cause	Remedy
1. OVERHEATING	Oil level too low.	Fill to proper level.
	Machine overloaded.	Reduce load.
	Plugged oil cooler core.	Clean core.
2. NOISY	Major failure or maladjustment.	Take machine to service shop for repair or adjustment.
3. OIL LEAKS	Loose bolts or damaged gaskets.	Tighten bolts. Have service shop replace gaskets.
	Loose fittings or oil lines.	Tighten fittings and lines.
4. MACHINE LACKS POWER OR ACCELERATION	Major failure or maladjustment.	Take machine to service shop for repair or adjustment.

POWER TRAIN (HYDROSTATIC TRANSMISSION)

Problem	Cause	Remedy
1. MACHINE WILL NOT MOVE	System low on oil.	Fill to proper level.
	Faulty control linkage.	Free linkage.
	Disconnected oil line.	Reconnect oil line.
	Mechanical failure.	See dealer for repair.
2. NEUTRAL HARD TO FIND	Faulty speed control linkage.	Adjust control linkage.
3. SYSTEM OVERHEATING	Oil level too low.	Fill to proper level.
	Oil cooler core plugged.	Clean core.
	Engine fan belt slipping or broken.	Tighten or replace belt.
4. SYSTEM NOISY	Air in system.	Check oil supply; fill if low. Check for loose fittings and tighten.
5. SLUGGISH ACCELER-ATION AND DECELERATION	Air in system.	Check oil supply; fill if low. Check for loose fittings and tighten.
6. HARD SHIFTING	Speed control lever not in neutral.	Position lever in neutral.

POWER TRAIN (DIFFERENTIAL)

Problem	Cause	Remedy
1. NOISY	Oil level too low.	Fill to proper level.
2. TURNING DIFFICULTY	Differential lock won't release, Brakes dragging, or differential lock stuck.	Take machine to service shop for repairs.
3. MECHANICAL LOCK DOESN'T HOLD	Linkage not in adjustment.	Take machine to service shop for repairs or adjustment.
4. HYDRAULIC LOCK DOESN'T HOLD	Valve malfunction or linkage not in adjustment.	Take machine to service shop for repairs or adjustment.

HYDRAULIC SYSTEM

Problem	Cause	Remedy
1. SYSTEM DOESN'T OPERATE	Little or no oil in system.	Fill to proper level. Check system for leaks.
	Oil of wrong viscosity.	Drain and refill system with proper oil.
	Oil filter plugged.	Replace filter.
	Restriction in system.	Take machine to service shop for repairs.
	Oil leaks.	Tighten fittings and lines.
	Slipping or broken pump drive belt.	Tighten or replace belt.
2. SYSTEM OPERATES ERRATICALLY	Air in system.	Check for leaks and tighten fittings and lines.
	Cold oil.	Allow system to warm up.

(continued on next page)

HYDRAULIC SYSTEM — Continued

Problem	Cause	Remedy
3. SYSTEM OPERATES SLOWLY	Cold oil.	Allow system to warm up.
	Oil viscosity too heavy.	Drain and refill system with proper viscosity of oil.
	Engine speed too slow.	Operate engine at recommended speed.
	Low oil supply.	Fill to proper level.
	Air in system.	Check for leaks and tighten fittings and lines.
4. SYSTEM OPERATES TOO FAST	Adjust cylinder ram speed. Adjust or repair.	See operator's manual. Adjust per operator's manual. For repair, see dealer.
5. SYSTEM OVERHEATING	Operator holding controls in power position too long.	Return controls to neutral when not in use.
	Incorrect oil viscosity.	Use recommended viscosity of oil.
	Low oil level.	Fill to proper level. Check for leaks.
	Dirty oil.	Drain and refill with clean oil.
	Oil cooler core plugged or dirty.	Clean core.
6. FOAMING OIL	Low oil level.	Fill to proper level. Check for leaks.
	Water in oil.	Drain and replace oil.
	Wrong kind of oil.	Drain and replace with recommended oil.
	Air leak.	Tighten lines and fittings
7. NOISY PUMP	Low oil level.	Fill to proper level. Check for leaks.
	Air in oil.	Check for leaks. Tighten lines and fittings.
8. COMPONENTS LEAKING	Failure of major parts.	Take machine to service shop for repairs.

BRAKES

Problem	Cause	Remedy
1. BRAKES NOT HOLDING	Glazed, greasy or worn linings.	Replace linings.
2. BRAKES NOT RELEASING	Cables or linkage binding.	Adjust linkage and cables.
	Foreign material lodged in brake mechanism.	Remove material.
3. PEDAL BOUNCES OR SPONGY	Hydraulic brakes: air in system.	Bleed brakes.
4. NO BRAKES	Hydraulic brakes: air in system.	Bleed brakes.
	Manual brakes: linings worn or linkage out of adjustment.	Replace linings or adjust linkage.
	Power brakes: accumulator discharged.	Take machine to service shop for repairs.
5. MACHINE PULLS TO ONE SIDE	Brakes adjusted unevenly.	Adjust brakes.
6. HYDRAULIC BRAKES OPERATE ERRATICALLY	Contaminated fluid.	Drain and clean system. Refill system with proper fluid. Bleed brakes.

OPERATOR'S CAB

Problem	Cause	Remedy
1. BLOWER NOT KEEPING DUST OUT	Defective seal around filter.	Check seal condition.
		Check filter for proper installation.
	Defective or dirty filter.	Replace or clean filter.
	Air leaks into cab.	Check cab for leaks. Plug leaks.
	Blower air flow too low.	See "BLOWER AIR FLOW TOO LOW."
2. BLOWER AIR FLOW TOO LOW	Clogged filter or air intake screen.	Clean filter or intake screen.
3. HEATER WILL NOT HEAT	Air trapped in heater core.	Bleed air from heater.
	Defective thermostat in engine.	Replace thermostat.
	Air conditioner turned on.	Turn off air conditioner.
4. HEATER WILL NOT SHUT OFF	Heater hoses connected improperly.	Change hose connections.
5. AIR CONDITIONER WILL NOT COOL	Blower air flow too low.	See "BLOWER AIR FLOW TOO LOW."
	Compressor belt slipping.	Tighten belt.
	Lack of refrigerant in system.	Check sight glass and have system recharged.
	Evaporator core plugged.	Clean core.
	Condenser core clogged.	Clean core.
	Heater turned on.	Turn off heater.
	Compressor not running.	Check fuse.

STORING MACHINES

When your machine is not going to be used for several months, it must be prepared for storage to prevent damage to components. Some manufacturers offer storage kits for long periods of storage. The kits may include rust and corrosion preventives and plastic bags and tape to seal off openings.

For suggested storage procedures, use the chart that follows. It tells what steps to perform to prepare your machine for storage and remove it from storage. Be sure to observe all safety precautions.

PREPARING MACHINE FOR STORAGE

*Step
No.*   *Operation*

1. **Clean exterior of machine.**

2. **Perform engine tune-up.**

3. **Check coolant for antifreeze protection to minimum anticipated temperature.**

   For those who don't drain and flush each season.

4. **Drain and refill transmission** *thoroughly. Heat before draining oil.*

5. **After refilling transmission, add corrosion and rust inhibitor to oil,** *if recommended.* **Operate transmission until oil is thoroughly circulated.**

6. **Clean and repack wheel bearings.**

7. **Drain and refill hydraulic system with fresh oil.**

NOTE: Operate hydraulic system until oil is thoroughly heated before draining.

8. **After refilling hydraulic system, add corrosion and rust inhibitor to oil,** *if recommended.* **Operate the system until oil is thoroughly circulated.**

9. **Park machine in selected storage location.**

10. **Drain fuel tank (gasoline and diesel only.)**

     CAUTION: Check with LP-Gas dealer about emptying LP-Gas fuel tank.

11. **Remove, clean and replace fuel sediment bowl and filters.**

12. **Add two gallons of fuel (mixed with rust inhibitor, if recommended) to the fuel tank.**

13. **Run engine for several minutes and then drain tank again.**

14. **Drain fuel lines and carburetor.**

15. **If recommended, add rust inhibitor to engine crankcase and to air intake.**

16. **With plastic bags and tape, seal the ends of air inlet pipe, exhaust pipe, crankcase breather pipe, and hydraulic system breather pipe.**

17. **Remove battery. (Check electrolyte level and specific gravity each month while the battery is in storage. Charge battery when necessary.)**

 CAUTION: Store battery where it can not be reached by children.

(continued on next page)

Step No.	Operation			

18. Remove any weights from machine.

19. Remove weights from tires. (Drain water if there's danger of freezing.)

20. Remove tires.

   If you do not wish to remove the tires, raise machine so that the tires are off the ground.

21. Check and inflate tires to normal pressures.

   NOTE: Support machine securely with support stands or blocks.

22. On machines with conventional dry clutch, block clutch pedal in the disengaged position.

23. Release tension on all drive belts and chains.

24. Apply grease or rust preventive to chains.

25. Coat all exposed metal surfaces, such as axles and hydraulic piston rods, with grease or a rust and corrosion preventive.

26. Lubricate all points normally requiring lubrication.

27. Check machine over carefully and make a list of parts or repairs needed to make the machine ready for operation when removed from storage. Make these repairs as soon as possible so that little time will be wasted when you remove the machine from storage to use during the next season.

28. Clean off rust, then prime and paint these areas to prevent further rusting.

29. Cover machine with a tarpaulin if it is not being stored in a building.

REMOVING MACHINE FROM STORAGE

Step No.   *Operation*

1. Remove all protective coverings from the machine, including plastic bags on all openings.

2. Remove clutch pedal block and allow clutch to engage.

3. Check tire inflation pressures.

4. Install tires on machine if they were removed.

5. Remove supports from machine and lower machine to ground.

6. Install battery. Check specific gravity of electrolyte and add water if necessary. Charge battery if specific gravity too low.

7. Check crankcase oil level.

   NOTE: Check operator's manual to see if machine can be operated with rust inhibitor in systems. If not, drain and refill.

8. Check hydraulic system oil level.

9. Check oil level in transmission, differential, and final drives.

10. Adjust tension of belt and chain drives.

11. Fill fuel tank with proper fuel.

12. Check coolant level.

13. Start the machine and let it idle for a few minutes. Check to be sure the machine is operating properly before using it.

GLOSSARY

A

ABRASION — Wearing or rubbing away.

ACCUMULATOR — A container which stores fluids.

ADDITIVE — An extra ingredient.

AGITATION — Mixing.

AIR CLEANER — A device for filtering dust from air.

AIR CONDITIONING — Control of temperature.

ALIGNMENT — Adjustment so wheels, pulleys, and gears are straight.

ALTERNATOR — A device which converts mechanical energy into electrical energy.

AMMETER — An instrument for measuring the *flow* of electrical current in amperes.

ANTIFREEZE — A material such as ethylene glycol, alcohol, etc., added to water to lower its freezing point.

AUTOMATIC — Without human action.

AUTOMATIC TRANSMISSION — A transmission that shifts gears by Itself.

AXLE — The shaft wheels are mounted on.

B

BACKFIRE — Ignition of the fuel and air mixture in the intake manifold of an engine.

BACKLASH — The clearance or ''play'' between two parts, such as meshed gears.

BALLAST — Weight.

BAR — A unit of pressure equal to 100 kiloPascals or 1.02 kg/cm². Approximately equal to 14.5 psi.

BEARING — The supporting part which reduces friction between two parts.

BELT TIGHTENER — The adjustable bolts and bracket assemblies that hold belt-driven parts, such as alternators, in position so the v-belt is tight enough.

BLEED — When air is removed from a fuel or hydraulic system.

BLOW-BY — When gas squeezes between the piston ring and the cylinder wall.

BOILING POINT — The temperature at which bubbles or vapors rise to the surface of a liquid and escape.

BREAK-IN — The process of wearing in to a desirable fit between the surfaces of two new or reconditioned parts.

BREAKDOWN — Same as break or fail.

BYPASS — An alternate path.

C

CAMSHAFT — A shaft with lobes that turns to open and close engine valves.

CARBON MONOXIDE — A poison gas produced by engines.

CARBURETOR — A device that mixes fuel with air.

CENTIMETER — A unit of length equal to 0.01 meter or approximately 0.39 inches.

CETANE — Measure of ignition quality of diesel fuel.

CHARGE — To restore the active materials in a battery by passing direct current through it.

CHATTERING — Vibrating.

CHOKE — A device placed in a carburetor air inlet to restrict the volume of incoming air.

CIRCUIT BREAKER — A device to protect an electrical circuit from overloads.

CLEARANCE — The space allowed between two parts, such as between a shaft and its bearing.

CLUTCH — A device for connecting and disconnecting the engine from the transmission.

COIL (IGNITION) — A transformer which increases voltage to produce strong ignition sparks.

COMBUSTION — The process of burning.

COMBUSTION CHAMBER — The space above the piston.

COMPRESSION — Squeezing.

COMPRESSION RATIO — The volume of the combustion chamber at the top of the compression stroke compared to the volume with the piston on bottom.

CONDENSATION — Process of changing a gas to a liquid.

CONDUCTOR — A substance an electrical current will flow through.

CONSTANT MESH TRANSMISSION — A transmission with the gears engaged at all times.

CONTRACTION — Shrinking.

CONVERTER — As used with LP-Gas, a device which changes LP-Gas from a liquid to a vapor for use by the engine.

COOLANT — A liquid circulated through an engine to cool it. Usually a mixture of water and antifreeze or rust inhibitor.

COUPLING — A connection between one part and another; may be mechanical, hydraulic, or electrical.

CRANKCASE — The lower housing of an engine which holds oil.

CRANKSHAFT — The main drive shaft of an engine.

CUSTOMARY UNITS — Measurements other than SI metrics measurements.

CURRENT — Movement of electricity through a conductor.

CYLINDER — A round container which holds a piston. Also a hydraulic power device.

CYLINDER HEAD — A detachable portion of an engine fastened on top of the cylinder block, which contains all or a portion of the combustion chamber.

D

DEAD CENTER — The extreme bottom or top position an engine piston can travel.

DETERGENT — A soap-like compound.

DETONATION — Explosion.

DIESEL ENGINE — Named after its developer, Dr. Rudolph Diesel. This engine ignites fuel in the cylinder from the heat generated by compression. The fuel is an "oil" rather than gasoline (petrol), and no spark plug or carburetor is required.

DIFFERENTIAL GEAR — The gear system which permits one drive wheel to turn faster than the other.

DIRECT DRIVE — Direct engagement between the engine and drive shaft where the engine crankshaft and the drive shaft turn at the same rpm.

DISCHARGE — To remove electrical energy from a charged body, such as a battery.

DISPLACEMENT — The volume displaced by the cylinders of a pump, motor, or engine. Also called swept volume.

DISTRIBUTOR (IGNITION) — A device which directs electricity to spark plugs.

DRAWBAR HORSEPOWER (Kilowatts) — Measure of the pulling power of a machine at the drawbar hitch point.

DRIVE LINE — The universal joints, drive shaft and other parts connecting the transmission to the driving wheels.

DRIVE SHAFT—Carries power from transmission to differential.

DYNAMOMETER — A test unit for measuring the actual power produced by an engine.

E

ELECTROLYTE — The sulfuric acid and water solution in a storage battery.

ENDPLAY — The amount of movement in a shaft due to clearance in the bearings.

ENGINE — The source of power in machines.

ENGINE DISPLACEMENT — The volume of an engine's cylinders.

EVAPORATION — Changing from a liquid to a vapor, such as boiling water to produce steam. Evaporation is the opposite of condensation.

EXHAUST — Gas which leaves the engine after combustion.

EXHAUST MANIFOLD — The passages from the engine cylinders to the exhaust pipe which conduct exhaust gases away from the engine.

EXPANSION — An increase in size. For example, when a metal rod is heated it increases in length and diameter.

F

FEELER GAUGE — A metal strip or blade of exact thickness.

FILTER — A device which removes solids from air or fluids.

FLOW METER — A hydraulic testing device which gauges flow rate.

FLOTATION — Ability of tires to stay on top of the ground.

FUEL KNOCK — Improper fuel burning.

FULL THROTTLE — Throttle open as far as it will go.

FUSE — A replaceable safety device for an electrical circuit. A fuse consists of a fine wire or a thin metal strip encased in glass or some fire-resistant material. When an overload occurs in the circuit, the wire or metal strip melts, breaking the circuit.

G

GALLON — 3.785 liters.

GAS — A substance which can be changed in volume and shape according to the temperature and pressure applied to it. For example, air is a gas which can be compressed into smaller volume and into any shape desired by pressure. It can also be expanded by the application of heat.

GASKET — A spacer between two parts. Also a seal.

GEAR — A cylinder- or cone-shaped part with teeth on one surface which mate with and engage the teeth of another gear.

GEAR RATIO — The ratio of the number of teeth on the larger gear to the number of teeth on the smaller gear.

GENERATOR — A device which converts mechanical energy into electrical energy.

GOVERNOR — A device to control and regulate engine speed. May be mechanical, hydraulic, or electrical.

GROUND — A ground occurs when any part of a wiring circuit touches a metallic part of the machine frame.

H

HECTARE — A measurement of area equal to 10,000 square meters, or approximately 2.47 acres (symbol: ha).

HORSEPOWER (hp) — The energy required to lift 550 pounds one foot in one second; approximately 0.746 kilowatts.

HYDRAULICS — Having to do with hydraulic power.

HYDRAULIC PRESSURE — Pressure exerted through liquid.

I

IDLE — Refers to the engine operating at its slowest speed.

INFLATE — Add air.

INHIBITOR — A material to restrain some unwanted action, such as a rust inhibitor in cooling systems.

INJECT — Spray in under pressure.

INJECTION PUMP (Diesel) — A device which meters fuel and delivers it under pressure to the engine injector.

INJECTOR (Diesel) — An assembly which receives a metered charge of fuel from the pump and injects the charge of fuel into a cylinder or chamber at high pressure.

INSULATOR — A substance or body that resists the flow of electrical current.

INTAKE — To bring in, or to draw in.

INTAKE MANIFOLD — The passages which bring in the fuel-air mixture from the carburetor to the engine cylinders.

INTAKE VALVE — A valve which permits fluid and gas to enter a chamber.

INTERLOCK — Mesh together.

INTERNAL COMBUSTION — Burning fuel in an enclosed space.

ISO — International Standards Organization

J

JOULE — A unit of energy or work approximately equal to 0.102 kgm, or about 0.7377 lb-ft (symbol: J).

K

KILOGRAM — A unit of mass (weight) approximately equal to 2.22 pounds (symbol: kg).

KILOMETER — A unit of distance equal to 1000 meters, or approximately 0.62 miles (symbol: km).

KILOPASCAL — A unit of pressure equal to 0.01 bar, or approximately 0.145 psi (symbol: kPa).

KILOWATT — A measurement of work equal to 1000 watts, or approximately 1.34 horsepower (symbol: kW).

KNOCK — A term used to describe engine noises made by improper ignition, or damaged parts.

L

LINKAGE — Any series of rods, yokes, and levers, used to transmit motion from one unit to another.

LITER — A unit of volume equal to 1000 cubic centimeters, or approximately 0.26 gallons (symbol: L).

LP-GAS, LIQUEFIED PETROLEUM GAS (LPG) — A fuel for internal combustion engines.

LUBRICATE — Applying grease, oil or other lubricant to make slippery.

M

MAINTENANCE — The care you give machines.

MANIFOLD — A pipe or casting used to carry a gas or liquid; an air intake manifold.

MANOMETER — A device for measuring a vacuum.

MESHING — Fitting together.

MECHANICAL EFFICIENCY (Engine) — The ratio between the indicated horsepower and the brake horsepower of an engine.

METER — A unit of length equal to 100 centimeters, or approximately 39.4 inches (symbol: m).

MISFIRING — Failure of an explosion to occur in cylinders while the engine is running.

MOMENTUM — Speed.

MOTOR — This term should be used for electric motors and should not be used when referring to an internal combustion *engine*.

MUFFLER — A chamber in the exhaust pipe which allows exhaust gases to expand and cool. It is usually fitted with baffles or porous plates and serves to reduce much of the noise created by the exhaust.

N

NEWTON — A unit of force equal to approximately 9.8 kilograms, or 4.45 pounds force (symbol: N).

O

OCTANE — Measurement of quality for gasoline.

OIL COOLER — A heat exchanger which removes heat from oil.

OHMMETER — An instrument for measuring the *resistance* in ohms of an electrical circuit.

OPEN CIRCUIT — An open circuit occurs when a circuit is broken.

OPERATOR'S MANUAL — Book written by a machine manufacturer telling how to use their product.

P

PARALLEL — Side by side.

PARTICLE — A tiny piece.

PINION — A small gear with teeth in the hub.

PISTON — A cylindrical part closed at one end, which is connected to the crankshaft by the connecting rod and pin. A cylindrical part inside a cylinder that is connected to a rod which transmits piston movement.

PISTON RING — An expanding ring placed in the grooves of the piston to seal it against the cylinder.

POLARITY — A term applied to the positive (+) and negative (−) ends of a magnet or electrical mechanism such as a coil or battery.

PORT — Passage or outlet.

POUND-FOOT (lb-ft) — This is a measure of the amount of energy or work required to lift one pound a distance of one foot; approximately equal to 1.36 joule.

POUR POINT — The lowest temperature that a fluid will flow.

POWER — Ability to do work.

PREIGNITION — Ignition occurring earlier than intended. For example, the explosive mixture being fired in a cylinder by a flake of incandescent carbon before the electric spark occurs.

PRELOAD — Pressure put on a bearing or bolt to eliminate any free play.

PRESSURE — Force, usually expressed in pounds per square inch (psi), kilopascals (kPa), or bar.

PUMP — A mechanical device for moving gas or fluid.

R

RATED HORSEPOWER — Value used by the engine manufacturer to rate the power of his engine, allowing for safe loads.

RATIO — The relation or proportion of one number to another.

REGULATOR (VOLTAGE) — A device which controls the flow of current or voltage in a circuit to a certain desired amount.

RELAY — An electrical switch which opens and closes a circuit automatically.

RESTRICTION — A blockage or plug.

RESERVOIR — A container for keeping a supply of fluid or gas.

S

SCALE — A flaky deposit occurring on steel or iron. Ordinarily used to describe the accumulation of minerals and metals accumulating in an engine cooling system.

SENDING UNIT — A device, usually located in some part of an engine or other system, to transmit information to a gauge on an instrument panel.

SHIM — Thin sheets used as spacers between two parts, such as the two halves of a bearing.

SHORT (OR SHORT CIRCUIT) — This occurs when one part of a circuit comes in contact with another part of the same circuit, diverting the flow of current from its desired path.

SLIDING GEAR TRANSMISSION — A transmission in which gears are moved on their shafts to change gear ratios.

SLUDGE — A composition of oxidized petroleum products along with an emulsion formed by the mixture of oil and water. This forms a pasty substance and clogs oil lines and passages and interferes with engine lubrication.

SOLVENT — A solution which dissolves some other material. For example, water is a solvent for sugar.

SPARK PLUGS — Devices which ignite the fuel by a spark in a spark-ignition engine cylinder.

SPECIFIC GRAVITY — The ratio of a weight of any volume of a substance to the weight of an equal volume of some substance taken as a standard, usually water for solids and liquids.

STORAGE BATTERY — A group of electrochemical cells connected together to store electricity.

STRAINER — A coarse filter.

STROKE — The distance moved by the piston in an engine or hydraulic cylinder.

SUCTION — Suction exists in a vessel when the pressure is lower than the atmospheric pressure; also see Vacuum.

SULFATION — The formation of hard crystals of lead sulfate on battery plates.

SYNCHROMESH TRANSMISSION — Transmission gearing which aids the meshing of two gears by turning both gears at the same speed.

SYNCHRONIZE — To cause two events to occur at the same time or in a proper sequence.

SYSTEM — A group of parts that work together.

T

TACHOMETER — An instrument for measuring rotary speed; usually revolutions per minute.

TENSION — Tightness.

THERMOSTAT — A heat-controlled valve used in air conditioning and heating, and in the cooling system of an engine to regulate the flow of water between the cylinder block and the radiator. Temperature control device.

TORQUE — Twisting or turning.

TORQUE CONVERTER — A turbine device to transmit power to a driven member by hydraulic action. It provides varying drive ratios; with a speed reduction, it increases torque.

TORQUE WRENCH — A special wrench with a built-in indicator to measure the applied turning force or torque.

TRANSMISSION — An assembly of gears, or other elements which gives variations in speed or direction between the input and output shafts.

TROUBLESHOOTING — A process of diagnosing the source of the trouble or troubles through observation and testing.

TUNE-UP — A process of accurate and careful adjustments to obtain the best performance from a machine.

TURBOCHARGER — A blower which forces air into the engine cylinders to produce more power.

V

VACUUM — A lack of pressure, a suction.

VACUUM GAUGE — An instrument designed to measure the degree of vacuum in a chamber.

VALVE — A device which controls either 1) pressure, 2) direction of flow, or 3) rate of flow.

VALVE CLEARANCE — The air gap allowed to compensate for expansion due to heat.

VAPOR LOCK — When the fuel boils in the engine fuel system, forming bubbles which retard or stop the flow of fuel to the carburetor.

VENT — An air hole.

VENTILATION — Fresh air.

VISCOSITY — The measure of thickness, resistance of a fluid to flow. An oil rating.

VOLATILITY — The tendency for a fluid to evaporate rapidly. For example, gasoline (petrol) is more volatile than kerosene as it evaporates at a lower temperature.

VOLTAGE — Force generated to cause current to flow in an electrical circuit. It is also referred to as electromotive force or electrical potential. Voltage is measured in volts.

VOLTMETER — An instrument for measuring the force in volts of an electrical current.

VOLUME — The amount of fluid flow in a certain time period. Usually given as gallons per minute (gpm), or liters per minute (L/min).

W

WARP — Bend and twist.

WATT — The rate of work done by an electric current of one ampere under a pressure of one volt (symbol: W).

WEAR — A reduction in size and strength.

WIRING HARNESS — A bundle of electrical wires.

ABBREVIATIONS

API — American Petroleum Institute (develops designations for various petroleum products)

ASAE — American Society of Agricultural Engineers (prepares standards for many agricultural machines and components.)

BDS — Bottom dead center (of an engine piston).

BTU — Abbreviation for *British Thermal Unit*. Amount of heat required to raise temperature of one pound of water (approximately one pint) 1° F. All substances are rated in relation to water as standard of measurement.

°C — degrees Celsius (of temperature) (formerly called centigrade)

cm — centimeter (length)

cm² — square centimeters (area)

cu. in. — cubic inch (volume)

°F — degrees Fahrenheit (of temperature)

ft. — foot or feet

gpm — gallons per minute (of fluid flow)

ha — hectare (area)

hp — horsepower (work or power)

I.D. — inside diameter (as of a hose or tube)

in. — inch (length)

ISO — International Standards Organization

J — joule (energy)

kg — kilogram (mass or weight)

km — kilometer (distance)

kPa — kilopascal (pressure)

kW — kilowatt (work or power)

L — liter (volume)

lb-ft — pound-foot (of torque or turning effort)

m — meter (length)

mph — miles per hour

N — Newton (force)

N•m — Newton•meter (torque)

mph — miles per hour

O.D. — outside diameter (as of a hose or tube)

psi — pounds per square inch (of pressure)

rpm — revolutions per minute

SAE — Society of Automotive Engineers (prepares standards for many machines and components for both automotive and agricultural use)

SI — International System of Units.

sq. ft. — square feet (area)

sq. in. — square inch (area)

TDC — top dead center (of an engine piston)

W — watt (work or power)

SUGGESTED READINGS

TEXTS

Farm Gas Engines and Tractors; Jones, Fred R.; McGraw-Hill Book Company, Inc., New York, 1963.

Farm Tractors; American Oil Company, Engineering Bulletin No. FT-53A, 910 South Michigan Ave., Chicago, Illinois 60680.

Farm Tractor Maintenance; Brown, Arlene D. and Morrison, Ivan G.; The Interstate Printers and Publishers, Danville, Illinois, 1958.

Implement and Tractor Shop Service; Technical Publications, Inc.; Kansas City, Missouri, 1987.

Tractors and Their Power Units; Barger, E.L.; Liljedahl, J.B.; Carleton, W.M.; and McKibben, E.G.; John Wiley and Sons, Inc., New York, 1963.

Tractor Maintenance Principles and Procedures; American Association for Vocational Instruction Materials, Engineering Center, Athens, Georgia 30601.

Farm Tractor Tune-up and Service Guide; American Association for Agricultural Engineering and Vocational Agriculture, Agricultural Engineering Center, Athens, Georgia 30601.

Fundamentals of Semiconductors; Delco Remy, Anderson, Indiana.

Transistor Regulators; Delco Remy, Anderson, Indiana.

How to Get Extra Service From Farm Tires; The Rubber Manufacturers Association, Inc., 444 Madison Avenue, New York.

The Engine Cooling System; National Carbon Company, 30 East 42nd Street, New York 17, New York, 1960.

Tractor Tips; Champion Spark Plug Company, Toledo, Ohio.

Fundamentals of Service: Engines; John Deere Service Training, Dept. F., John Deere Road, Moline, Illinois 61265.

Fundamentals of Service: Electrical Systems; John Deere Service Training, Dept. F., John Deere Road, Moline, Illinois 61265.

Fundamentals of Service: Hydraulics; John Deere Service Training, Dept. F., John Deere Road, Moline, Illinois 61265.

Fundamentals of Service: Power Trains; John Deere Service Training, Dept. F., John Deere Road, Moline, Illinois 61265.

Fundamentals of Service: Air Conditioning; John Deere Service Training, Dept. F., John Deere Road, Moline, Illinois 61265.

Fundamentals of Service: Fuels, Lubricants and Coolants; John Deere Service Training, Dept. F., John Deere Road, Moline, Illinois 61265.

Fundamentals of Service: Tires and Tracks; John Deere Service Training, Dept. F., John Deere Road, Moline, Illinois 61265.

Fundamentals of Service: Belts and Chains; John Deere Service Training, Dept. F., John Deere Road, Moline, Illinois 61265.

Fundamentals of Service: Bearings and Seals; John Deere Service Training, Dept. F., John Deere Road, Moline, Illinois 61265.

Fundamentals of Service: Shop Tools; John Deere Service Training, Dept. F., John Deere Road, Moline, Illinois 61265.

Fundamentals of Service: Identification of Parts Failures; John Deere Service Training, Dept. F., John Deere Road, Moline, Illinois 61265.

Fundamentals of Service: Fiber Glass and Plastics; John Deere Service Training, Dept. F., John Deere Road, Moline, Illinois 61265.

Fundamentals of Machine Operation: Tractors; John Deere Service Training, Dept. F., John Deere Road, Moline, Illinois 61265.

Fundamentals of Machine Operation: Combine Harvesting; John Deere Service Training, Dept. F., John Deere Road, Moline, Illinois 61265.

Fundamentals of Machine Operation: Hay and Forage Harvesting; John Deere Service Training, Dept. F., John Deere Road, Moline, Illinois 61265.

Fundamentals of Machine Operation: Machinery Management; John Deere Service Training, Dept. F., John Deere Road, Moline, Illinois 61265.

Fundamentals of Machine Operation: Agricultural Machinery Safety; John Deere Service Training, Dept. F., John Deere Road, Moline, Illinois 61265.

Conservation Farming; John Deere Service Training, Dept. F., John Deere Road, Moline, Illinois 61265.

INSTRUCTOR'S GUIDE

Machinery Maintenance Instructor's Guide. Contains activities, exercises, and transparency masters based on text. John Deere Service Publications, Dept. 333, John Deere Road, Moline, IL 61265.

MEASUREMENT CONVERSION CHART

Metric to English

LENGTH

1 millimeter = 0.039 37 inches ....in.
1 meter = 3,281 feet ....ft
1 kilometer = 0.621 miles ....mi

AREA

1 meter^2 = 10.76 feet^2 ....ft^2
1 hectare = 2.471 acres ....acre
   (hectare = 10,000 m^2)

MASS (WEIGHT)

1 kilogram = 2.205 pounds ....lb
1 tonne (1000 kg) = 1.102 short ton ....sh tn

VOLUME

1 meter^3 = 35,31 foot^3 ....ft^3
1 meter^3 = 1.308 yard^3 ....yd^3
1 meter^3 = 28.38 bushel ....bu
1 liter = 0.028 38 bushel ....bu
1 liter = 1.057 quart ....qt

PRESSURE

1 bar = 14.50 pound/in^2 (psi) ....psi
   (1 bar = 10^5 pascal)

STRESS

1 megapascal or
1 newton/millimeter^2 = 145 pound/in^2 (psi) ....psi
   (1 N/mm^2 = 1 MPa)

POWER

1 kilowatt = 1.341 horsepower (550 ftlb/s) ....hp
   1 watt = 1 Nm/s)

ENERGY (WORK)

1 joule = 0.000 947 8 British Thermal Unit ....BTU
   (1 J = 1 W s)

FORCE

1 newton = 0.2248 pounds force ....lb

TORQUE OR BENDING MOMENT

1 newton meter = 0.7376 pound-foot ....(lb-ft)

TEMPERATURE

$t_C = (t_F - 32)/1.8$

English to Metric

LENGTH

1 inch = 25.4 millimeters ....mm
1 foot = 0.3048 meters ....m
1 yard = .9144 meters ....m
1 mile = 1.608 kilometers ....km

AREA

1 foot^2 = 0.0929 meter^2 ....m^2
1 acre = 0.4047 hectare ....ha
   (hectare = 10,000 m^2)

MASS WEIGHT

1 pound = 0.4535 kilograms ....kg
1 ton (2000 lb) = 0.9071 tonnes ....t

VOLUME

1 foot^3 = 0.028 32 meter^3 ....m^3
1 yard^3 = 0.7646 meter^3 ....m^3
1 bushel = 0.035 24 meter^3 ....m^3
1 bushel = 35.24 liter L
1 quart = 0.9464 liter L
1 gallon = 3.785 liters L

PRESSURE

1 pound/in^2 (psi) = 0.068 95 bar ....bar
   (1 bar = 10^5 pascal)

STRESS

1 pound/in^2 (psi) = 0.006 895 megapascal ....MPa
                     or newton/mm^2 ....N/mm^2
   (1 N/mm^2 = 1 MPa)

POWER

1 horsepower (ft-lb/s) = .7457 kilowatt...........................kW
   (1 watt = 1 N·m/s)

ENERGY (WORK)

1 British Thermal Unit = 1055 joules ....J
   (1 J = 1 W s)

FORCE

1 pound = 4.448 newtons ....N

TORQUE OR BENDING MOMENT

1 pound-foot = 1.356 newton-meters ....N•m

TEMPERATURE

$t_F = 1.8\, t_C + 32$

INDEX